Screenformation 2.0

2nd Edition of the book that brings together major scientific studies, educational research, and expert commentary on the harmful effects of screentime

Robert Rose-Coutré, M.A.

Second Edition 2023
Second printing November 2024
ISBN (e-Book): 978-0-9973250-5-8
ISBN (Paperback): 978-0-9973250-6-5
ISBN (Hardcover): 978-0-9973250-7-2
Library of Congress Control Number (LCCN): 2018900932

Cover Design by Robert Rose-Coutré

Printed in the United States of America

West Chester, Pennsylvania
Rose-Coutré Publishing
2023, e-Book
2023, Paperback
2023, Hardcover

Table of Contents

Dedication

This book is dedicated to my wife, Mitra Rose-Coutré

Acknowledgement

I would like to acknowledge Mitra Rose-Coutré for her astute editorial contributions and advice.

PART I:
Screen Science

INTRODUCTION

Screentime has been a topic of controversy for several generations. We've heard how our digital world promotes global communications, provides learning opportunities that were not possible in the past, helps us stay in touch with each other long distance, and gives us access to almost any information instantly. We hear digital technology can bring the global human family closer together.

On the other hand, we've heard that screentime isolates us, divides us into our segregated echo chambers, leads to depression, impairs mental and physical development, and takes away from more active and social pastimes in the real world.

These are commonly discussed issues, but they are not commonly understood. One of the first misconceptions people have is that there's not a lot of research on the effects of screentime. The fact is, there is a wealth of scientific studies, educational research, and expert commentary on how screentime has impacted individuals and the population writ large today and over the past fifty years. But that rich scientific knowledge is not widely communicated to the general public; hence, the misconception.

To help remedy that misconception, this book compiles and communicates the rich body of knowledge to the reader. It provides a comprehensive, well-documented investigation into how mind and body respond to screentime. It pulls together the findings into one place for convenient access, including research from the biological, psychological, physiological, sociological, and behavioral sciences on the subject of screentime. The Works Cited section provides more than 200 references, which are explored in-depth throughout the text.

The findings are astounding in their consistency and universality, pointing to the same conclusion: Screentime significantly damages many areas of brain development, mental health, and physical health. To name a few areas impacted, screentime damages basic brain function, mental health, creativity, reasoning, concentration, attention span, problem solving, social skills, empathy, emotional development, motor skills, and physical health.

If that seems exaggerated, don't take our word for it. Check the studies, the research, the findings, and the facts, which are now readily available and easy to access.

Screen-related damage takes many forms. Brain and body suffer, quality of personal and social life degrades, overall human experience shrinks down to a fraction of what we experienced all those millennia before the age of screens. But the bad news conversely points to the better path forward. With heightened awareness of the harmful impacts, we can build our own immunity to screen-related injury. Instead of automatically following everyone else's daily slog through the mind-numbing sea of screens, we can forge our own active-living path, to create a better life for ourselves and our children.

CHAPTER 1: BRAIN AND BODY DEVELOPMENT

Alpha Waves

We begin our investigation into the science of screentime by addressing the screen's biological effect on brainwaves. Brainwaves are associated with states of mind, such as critically alert (Beta Waves) versus passively receptive (Alpha Waves).

Based on a groundbreaking "Alpha Wave" study by Dr. Herbert E. Krugman[1], and others since then, screen viewing alters the brain by making it suggestible. The act of watching content on-screen switches our brainwaves from active to passive.

Our awake state involves Beta waves. Beta waves facilitate active thinking, "beta waves are associated with alertness, activity." Conversely, Alpha waves are slow and are not associated with alertness. "Alpha waves are not simply slow Beta waves; they are a new parameter." Krugman's research revealed that screen viewing switches the brain from Beta to Alpha waves. Krugman explains how our brainwave variation makes us process screen information differently from print information. The print-reading response is "active and composed primarily of fast brain waves, whereas the response to television might be understood as passive and composed primarily of slow brain waves." Perhaps the most profound finding in the brainwave discovery was that the brain switches from active-thinking Beta Waves to passive non-thinking Alpha Waves in less than a minute from the start of viewing, and it makes no difference what's on—the screen itself causes the switch to passivity.[2]

Suggestibility makes us easier to influence. We are more receptive to the simulation of reality presented on the screen. That fact alone—that screen-viewing makes our brains less active and more easily influenced—undermines brain health in general. An abundance of subsequent biological and psychological research has shown that screentime causes mental decline, altered perception of reality, and other types of damage in addition to suggestibility.[3] It should be no surprise that Alpha waves are also associated with hypnosis and hypnotic suggestion. When we are watching our screens, we are not aware of this shift from active to passive brainwaves.

We watch TV approximately 30 hours a week.[4] On smartphones, tablets, and laptops, we are estimated to consume the equivalent of nine DVDs-worth of data per day, per person.[5] As reported by Victor C. Strasburger, MD, in an American Academy of Pediatrics study, "Young people now spend more time with media than they do in school—it is the leading activity for children and teenagers other than sleeping."[6] As a result, we spend almost every hour of every day in a passive Alpha Wave state of suggestibility.

This shift in our population from spending our waking hours in an active alert state of mind, to spending our waking hours in a passive suggestible state of mind, is a relatively recent shift. Humans spent 200,000 years in the active Beta Wave state, 100 percent of every waking hour, every day. Suddenly in the past 30 to 50 years, we became a perpetually passive, receptive species, limited to a lifelong Alpha Wave state of existence. Our new perpetual screen-surround life experience undermines human development, which in turn diminishes human variation. It increases our conformity and uniformity. The shift has a dramatic narrowing and homogenizing effect on our species.

Critical Processing

Screentime suppresses our ability to critically process information. Our minds are constantly bombarded by stimuli and data—news and information conveyed via fast-changing images on a screen. As we saw in the "Alpha Waves" section, the screen puts us into a passively receptive state. But before the mind can analyze news and information successfully, it must be in an active state. Then it can draw comparisons and relations with sustained comprehension.

Reading or talking to another person activates the region of the brain that prepares us to logically process information and critically evaluate it. The mind is already "in gear" as it encounters arguments and concepts in a book or in a conversation.[7] Conversely, watching TV or video does the opposite. In experiments tracking brainwaves while subjects watched TV, "The EEG studies similarly show less mental stimulation, as measured by alpha brainwave production, during viewing than during reading."[8]

We now know this fact: we cannot critically process video-format information as reliably as reading the same information. That means when we consume information via TV and video, we are more likely to accept whatever we see "as true" without thinking, and without realizing it (unless it contradicts our pre-existing *confirmation bias*, covered later in this book—in that case, we dismiss it out of hand as false, also without thinking). Watching video, our brain is not prepared to process anything competently.[9]

Even instructional videos—created and consumed for the purpose of learning—fail to instruct. Ed O'Brien, social psychologist at the University of Chicago, devised a series of experiments to understand the effect of instructional videos on people.[10] He found there was a gap between the perception of learning and actual learning. Volunteers watched a variety of instructional videos from one time to twenty times. Participants who watched twenty times were much more confident that they could perform what the video was teaching. In fact, none of the participants improved by watching the video. Whether watching once, or twenty times, "there was absolutely no effect of video-watching on actual performance." Even the most confident participants could not transfer watching, to actual learning or doing. Watching an instructional video, "It seems to convince people that they have learned everything to perform the same task." When they try to perform the task, however, they discover they did not learn anything. Conversely, the study showed "Reading an instruction manual … didn't produce the same overconfidence." As one of the participants noted, "When I watch the video, what I think I'm losing is, I'm losing the ability to gain the skill. I think I'm tricking my mind to think that, you know, I'm getting that skill. I watched the video. I know how to do it. In reality, that's not true."[11]

The overwhelming evidence from multitudes of studies shows us that screentime impairs our learning ability and damages our brain capabilities in many ways.[12] But it's not as harmless as merely becoming less informed. The screen shapes our minds to be vulnerable and receptive to simplistic screen-information, including *mis*information.

Video-induced Alpha Waves make us accept bad information while making us less competent to intelligently evaluate any information.[13] With impaired logical processing, our receptive condition renders us defenseless against manipulation, even while we think we are in full command of our faculties.

We cannot feel the Beta-Alpha "switch" happening while we are watching screens. That lack of awareness makes us more easily influenced. We think we are processing information just fine, but in fact we are accepting information without critically processing it.

The screen experience biologically prevents us from thinking for ourselves. A screen-informed mind is not an informed mind. It is merely reshaped and populated by prepackaged messaging. The decision to rely on TV for information is a poor decision. It illustrates the poor judgment brought about by watching TV. The vicious cycle feeds itself.

Secondhand Exposure

We have found ample evidence of the screen's debilitating effects on our mental abilities. But the effects on children are more insidious. The Alpha wave state undermines child development even when the children are not directly watching the screen. American Academy of Pediatrics Research on the effects of TV on children found that "Casual exposure (to TV) can harm their language development, making it harder for them to cope when they go to school ... Children are as vulnerable to the effects of 'passive TV' as they are to secondhand smoking, according to experts... As well as discouraging the amount of screen-time to which youngsters are exposed, it cautioned against adults watching television with them nearby. It said parents needed to understand that 'their own media use can have a negative effect on children'. ... the data should serve as a 'wake up call' to parents."[14]

The wakeup call to parents is especially important during the first two years of their child's life: "The risk of television delaying learning in infants is so great that the American Academy of Pediatrics recommends that babies under the age of 2 be banned from watching altogether."[15]

While the very atmosphere created by a TV in the room undermines child development, the atmosphere created by books in the room actually enhances child development. A study of 160,000 adults between 2011 and 2015 found that having eighty or more books in the home results in adults with stronger cognitive competencies. The study found that a home library promotes greater math and literacy skills, information communication technology (ICT) skills, and lifelong habits of knowledge seeking. Adults with high-school-level education acquire these skills equal to university graduates who grew up with only a few books in the home.[16] It's the parents' job to set the tone, better or worse for their children's future.

Mental Muscle

Like a muscle, the mind requires exercise to build strength. As a result of mental exercise, we become smarter. As *Emotional Intelligence 2.0* author and clinical psychologist Travis Bradberry notes, "your brain grows new connections much as your biceps might swell if you started curling heavy weights several times a week. The change is gradual and the weight becomes easier and easier to lift the longer you stick to your routine ... the brain cells develop new connections to speed the efficiency of thought…."[17]

By failing to exercise brain cells, they weaken and don't function as well. Screentime is not brain-exercise, it's brain-massage. Screen-stimulation acts like a delicious massage to the brain. The brain weakens like a muscle that is pleasantly massaged, but never exercised. Without exercise, the massaged mind feels good, while getting flabby. A multitude of studies and research have established the fact that passive screentime atrophies the mind. We know that screentime robs the mind of energy, thinking, problem-solving skills, logic, empathy, creativity, mental health, and emotional development.[18]

Replacing mental exercise with mental massage, people become less able to "self-stimulate" their own brains, or initiate their own mental activity, the way humans did from the beginning of human history until the mid-1900s. As the passive screen experience effortlessly stimulates brain cells, so the ability to initiate one's own brain-cell activity is put to sleep, deteriorated like an atrophied muscle.[19]

Children are especially vulnerable to damage from a lack of mental exercise. Noted in the *Scientific Learning* article entitled, "This is Your Child's Brain on TV": "Children require face-to-face contact from caretakers who provide verbal and non-verbal clues to kids that television—no matter how kid-friendly—cannot … Since brain circuits organize and reorganize themselves in response to an infant's interactions with his or her environment, exposing babies to a variety of positive experiences (such as talking, cuddling, reading, singing, and playing in different environments) not only helps tune babies in to the language of their culture, but it also builds a foundation for developing the attention, cognition, memory, social-emotional, language and literacy, and sensory and motor skills that will help them reach their potential later on."[20]

The important takeaway is that screentime cannot replace this foundational real-world development.

Motor Skills

According to child-development experts, children's mobility levels are at an all-time low. Research has revealed that a concerning number of today's 4-year-olds are not physically ready to start school. Screentime is preventing development of fundamental coordination such as balance and motor skills.

"They suffer lack of motor skills and reflexes. Almost 90 percent of children demonstrated some degree of movement difficulty for their age… Children lack the ability to complete simple tasks such as sitting still, holding a pencil, putting on their shoes, and especially reading… In a supplementary study of 25 Foundation Stage teachers, 80 percent said they had identified a sudden decline in physical mobility happening within the past three to six years…The reason? Today's children are less active in their early years compared with previous decades, with typical movements associated with play and development reduced by electronic toys and screens."[21]

More studies are finding the same results. A December 2018 clinical report from the American Academy of Pediatrics found that as screentime displaces traditional play and traditional toys (e.g., fine motor examples, such as blocks, shapes, puzzles, trains, etc., and gross motor examples, such as large toy cars, tricycles, and push and pull toys), children experience reductions in both fine motor, adaptive, and gross motor skills and abilities.[22]

One troubling effect of these childhood deficiencies is permanently reduced dexterity through adulthood. For example, as *The Daily Mail* reports, there has been "a decline in the dexterity of students in just the past decade … Surgery students struggle to use their hands because they spent too much time in front of a screen growing up." Surgeons are becoming less developed in tactile skills, "less competent and less confident in using their hands." This is not encouraging for those who will need delicate life-and-death surgery in the future. But the dexterity deficiency affects all professions, not just medicine. Think of any assembly-line factory and the replacement of humans with robots. As humans become less competent with their hands, robots increasingly replace people in the workplace.[23]

We can no longer assume people can do practical things, cutting things out, making things. If in this fundamental area of physical development, using our hands for basic activities, standards become lower across the population.

Verbal Skills

One of the most important areas of child development, linked to overall cognitive health, is verbal skill. Children watching TV suffer severe declines in vocabulary, which damages their ability to think. Their verbal skills are less developed than they would have been if they had not watched TV. Scientific studies have shown that children learn words by interacting verbally with other humans—by watching and listening to parents in real conversations—and that screentime delays verbal development.[24] TV cannot imitate it, replace it, or even supplement it. TV can only undermine it.

To fully learn words and concepts, we have to experience the moment when words are used in real life, in relation to real people and things in a three-dimensional-world context. TV experience is not a real-world experience. If we don't experience real-world context, we don't learn as well. "Research conducted during the next two decades removed any doubt about the impact of early brain stimulation on a child's later cognitive development. … they were able to demonstrate that environmental factors can alter neuron pathways during early childhood and long after … among the most important of the environmental factors … are the language and eye contact an infant is exposed to [and] … the number of words an infant hears each day is the single most important predictor of later intelligence, school success and social competence. But there's one catch. The words have to come from an attentive, engaged human being … radio and television do not work" (Winn).[25]

With real-life interaction, as with reading, the brain is activated and able to learn. The effect of TV is to turn off the learning process.

The American Academy of Pediatrics warns parents, "It may be tempting to put your infant or toddler in front of the television, especially to watch shows created just for children under age two. But the American Academy of Pediatrics says: Don't do it!"[26]

Comparisons

People have always been different—some better some worse—in every skill. But comparing self with others is disingenuous because it dodges the important comparison: self "as is" versus self "as could be."

Columbia University Professor of Psychology E. Tory Higgins explains that recognizing this discrepancy—"actual self" versus "can be" self—is referred to as self-discrepancy—"i.e., I am *not* fulfilling my potential."[27] It is known to be associated with depression: "What's wrong with me? I know I can be what I would like to be but I am not doing it."[28] It goes hand-in-hand with related sentiments about screen-viewing behaviors: "I don't want to watch as much as I do, but I can't help it."[29]

But recognizing the discrepancy between the "as is" self and the "as could be" or "I can be" self, is an essential step towards deciding to close the gap. Facing failed potential, the "dissatisfaction … may motivate someone to try harder. (e.g., '…Why am I so lazy? I know I can do better. I've got to try harder!')."[30]

Passive screentime makes our "as is" self *lower functioning* than our "as could be" self. Screentime also undermines our motivation so it becomes harder and harder to climb out of this rut. Screentime diminishes inherited potential, regardless of how one compares with others. Over time, lower functioning manifests itself as less achievement, and therefore less fulfillment. The saddest part of choosing a "lesser me" is the regret, which comes later in life when we realize what we have lost. As *The Motivation Myth* author Jeff Haden aptly states on the topic of this type of regret, "Sure, the work is hard. Sure, the work is painful—but it's significantly less painful than thinking back on what will never be."[31] The "as could be" self is now out of reach.

The old saying is true: "It's not about being better than someone else, it's about being better than you were the day before."

Effect on Intelligence

A recent study reported in an article entitled, "Watching lots of TV 'makes you stupid,'" confirms dozens of previous studies: screentime damages brain function.

- "The study found people who watch the most TV are twice as likely to have poor mental functioning.
- "Watching TV for hours impairs your mental ability, according to study.
- "It is one of the first studies to demonstrate that watching too much TV encourages cognitive aging, even before middle age."[32]

"Several previous studies have found lower verbal IQ and increased aggressiveness in proportion to the amount of television children watch. This new research uncovers the biological mechanism for these changes in behavior and *drop in intelligence*."[33] As clinical psychologist Dr. Azmaira H. Maker adds, screentime also shrinks the brain and impairs overall cognitive function.[34]

Supported by further research, child psychiatrist Dr. Victoria Dunckley explains that allowing children to be entertained by screens causes underdeveloped brains, atrophy, brain shrinkage, and severed brain connectivity. Screentime literally makes the brain smaller and less connected. It results in children less able to learn, concentrate, form relationships, regulate moods, and causes many other deficiencies in brain activity.[35]

The key to these findings is the mathematical certainty: as hours of watching TV goes up, intelligence goes down.

Similar results were found in a recent American Medical Association study. The study compared preschoolers who get a lot of screentime with preschoolers who get very little screentime. The preschoolers with a lot of screentime suffered disorganized executive functions and were underdeveloped in language and literacy skills. The deficiencies were shown both in behavioral testing and in MRI brain imaging. The MRIs show startling deficiencies in brain development. The lack of brain development also causes slower processing speed. "'This is the first study to document associations between higher screen use and lower measures of brain structure and skills in preschool-aged kids,' said lead author Dr. John Hutton, a pediatrician and clinical researcher at Cincinnati Children's Hospital." Conversely, reading, juggling, and learning a musical instrument improve the organization and structure of the brain (demonstrated both in behavioral testing and in brain imaging).[36]

The depth and complexity of our damage to brain function is worse than we might expect. Maryanne Wolf, Director of the Center for Reading and Language Research, and Professor of Child Study and Human Development at Tufts University, puts today's screen culture into perspective:

> "The iPad is the new pacifier for babies and toddlers. Younger school-aged children read stories on smartphones; older boys don't read at all, but hunch over video games. Parents and other passengers read on Kindles or skim a flotilla of email and news feeds. Unbeknownst to most of us, an invisible, game-changing transformation links everyone in this picture: the neuronal circuit that underlies the brain's ability to read is subtly, rapidly changing."[37]

Professor Wolf explains that our immersion in the new digital world is undermining "our most important intellectual and affective processes: internalized knowledge, analogical reasoning, and inference; perspective-taking and empathy; critical analysis and the generation of insight."

Dr. Wolf notes a "series of studies which indicate that the 'new norm' in reading is skimming"—thus the "subtle atrophy of critical analysis and empathy," abilities that require slow and deep reading for long periods at a time, in order to develop.

These deficiencies surface in college as "students no longer have the patience to read longer, denser, more difficult texts" and can no longer "read with a level of critical analysis sufficient to comprehend the complexity of thought and argument."[38]

Further widespread consequences include a decreasing ability in the population to tell the difference between logical or factual statements versus emotive statements that may be false but "ring true" due to strong emotional triggers. This is the inevitable result of a screen culture, and it is today's reality.

This reality is not confined to the US. A 2017 European study shows that "Intelligence levels are falling in advanced industrial countries across Europe."[39] Authors of the report James Flynn and Michael Shayer say the results point to a wider global trend. Perhaps most troubling, is that the worst losses we are now experiencing in the population is at the top. We are losing our best thinkers to the debilitating effects of screentime in our upbringing. As the study notes, "Looming over all is their message that the pool of those who reach the top level of cognitive performance is being decimated: fewer and fewer people attain the formal level at which they can think in terms of abstractions and develop their capacity for deductive logic and systematic planning."[40]

The Dunning-Kruger Effect

The less we know, the more we think we know. As we lose our ability to think, the higher our opinion of our ability to think. As the Dunning-Kruger Effect proved, the less we can logically process information, the more we grossly overestimate our ability to logically process information.[41]

The spiraling descent of competence gains momentum: lower brain function, higher confidence in brain function; worse decision making, greater certainty of good decision making. The Dunning-Kruger Effect—the collapse of accurate self-appraisal—spreads like an intelligence-cancer. We become less smart with every hour, and more convinced that we are smarter than ever.

As David Dunning comments about his Dunning-Kruger experiments, "A whole battery of studies conducted by myself and others have confirmed that people who don't know much about a given set of cognitive, technical, or social skills tend to grossly overestimate their prowess and performance, whether it's grammar, emotional intelligence, logical reasoning, firearm care and safety, debating, or financial knowledge. College students who hand in exams that will earn them Ds and Fs tend to think their efforts will be worthy of far higher grades; [they] overestimate their competence by a long shot." He highlights the irony: "In many cases, incompetence does not leave people disoriented, perplexed, or cautious. Instead, the incompetent are often blessed with an inappropriate confidence, buoyed by something that feels to them like knowledge."[42]

Confirmation Bias

Confirmation bias and the related tendency of self-segregation are two powerful forces that erode critical thinking. Screentime exacerbates both.

The increase of screentime with smartphones and social media has heightened our susceptibility to "close ranks" socially, a voluntary tendency towards self-segregation. Screenlife makes it easy to exclude people and comments that we don't already believe and love. As *The Death of Expertise* author Tom Nichols observes, "the Internet has politically and intellectually mired millions of Americans in their own biases. Social media outlets such as Facebook amplify this echo chamber." Even if we don't intentionally block what we disagree with, Facebook feeds us what we "like" as part of its service.[43] We segregate from differing people as well as from differing information sources. By replacing social life with social media, we reduce or remove exposure to differing views.

Relying on the Internet for research, news, and information reinforces our mistaken ideas and biases. As author Thomas Friedman says in his book *Thank You for Being Late*, "The Internet is an open sewer of untreated, unfiltered information."[44] There are no signposts to tell us which sites are truly peer-reviewed scientific sources, or rigorous high-caliber journalism; versus poorly substantiated "sciencey" sites, yellow journalism, or flagrant misinformation. Instead of searching for truth, we search to confirm our pre-existing opinions, auto-select sources we like, and hear what we want to hear.

The name for this behavior is *confirmation bias*. "Once we have formed a view, we embrace information that confirms that view while ignoring, or rejecting, information that casts doubt on it….We pick out those bits of data that make us feel good because they confirm our prejudices."[45] Confirmation bias is operating when we:

- Overlook bad arguments if they support our beliefs
- Ignore faulty logic if we feel good about the conclusions
- Don't bother to investigate an experiment if we agree with the results
- Accept what we like, and reject what we don't like, regardless of the truth.

The Internet, TV, and other screen-media promote confirmation bias, which further segregates society. By definition, confirmation bias means we can always feel right about everything we think. As a result, we stop growing.

"Today, there is a news source for every taste and political view, with the line between journalism and entertainment intentionally obscured to drive ratings and clicks." Furthermore, we choose our sources according to our bias, "people gravitate toward sources whose views they already share."[46] News channels must chase oversimplified political demographics to make money for their advertisers. The media promotes artificial divisions between us, which gradually become real divisions between us. We are molded by what we watch, we segregate ourselves accordingly.

Here we see the vicious cycle of screen-centric news and information:

- Oversimplified biased reporting that promotes hostility and division to woo the "liberal market" or the "conservative market" to generate revenue, instead of neutral objective reporting, which doesn't win ratings or generate revenue.
- Uncritical acceptance (if you already agree) and uncritical rejection (if you already disagree) of the message: confirmation bias.

- Self-perpetuated social segregation.

To use the congressional metaphor, both sides of the aisle are equally guilty of confirmation bias. Studies show that we all (conservatives and liberals alike) reject evidence that upsets our worldview, we "just don't believe it." To add insult to injury, our worldview today embraces screen devices and screentime as normal (a major cause of confirmation bias). Data showing that screentime damages our mental and physical health is neither welcomed nor heeded. Conservatives and liberals alike will deny science when it challenges their worldview.[47]

This trend will continue to worsen as long as we receive our information from screens. Watching the news on a screen makes us much more susceptible to confirmation bias. As demonstrated earlier, screen media disengages logical processing and critical thinking—we respond with a predictable emotional reflex instead of an intelligent analysis. If the message feels good, then it's true; if it feels bad, then it's false.

Effect on Knowledge Acquisition

Screen-based information gathering drives us further away from knowledge, and further into the illusion of knowledge. Reports showing the trend are increasingly common. "A 2011 University of Chicago study found that America's college graduates 'failed to make significant gains in critical thinking and complex reasoning during their four years of college,' but more worrisome, they 'also failed to develop dispositions associated with civic engagement.' … they were less interested in applying what little they might have learned to their responsibilities as citizens."[48]

Younger people are especially susceptible to the tendency to read uncritically when reading on the Internet. The University College of London (UCL) study found that they typically skim and are not "evaluating information, either for relevance, accuracy or authority. … Internet users tend to … believe whichever results of a search come up first in the rankings, mostly without regard to the origins of those results."

Increasingly, people believe what they see, without regard for the source, as long as it has high search ranking. Clever search-engine optimization (SEO) now trumps knowledge and truth in the minds of people "reading" on the Internet. The UCL study authors concluded that society is "dumbing down." But it gets worse. Experimental psychologists at Yale found that "'people who search for information on the Web emerge from the process with an inflated sense of how much they know' … This is a kind of electronic version of the Dunning-Kruger Effect, in which the least competent people surfing the web are the least likely to realize that they're not learning anything."[49]

The power of "mindless confidence" on a topic should not be underestimated. Even faced with hard evidence that they are incorrect, people with the illusion of knowledge simply "doubledown" on their false assumptions. People respond to truth by holding tighter to falsehood despite clear indications that they're wrong. This is the increasingly common "backfire effect" when trying to inform someone of something they don't want to know.[50]

Embracing falsehood to protect our feelings is no respecter of political orientation. As recent studies make abundantly clear, "when exposed to scientific research that challenged their views, both liberals and conservatives reacted by doubting the science, rather than themselves." Increasingly people will cling to a comfortable sense of rightness and reject the evidence.[51] (See also the "The Dunning-Kruger Effect" and "Confirmation Bias" sections.)

Our society is fast becoming a prototype Dunning-Kruger experiment, where knowledge is shallow or nonexistent, while our confidence in our rightness is off-the-charts more exaggerated than ever in history. We are raising our children to make every next generation worse than the last. We think we are preparing our children for "today's technological society" by giving them more Internet time in schools and at home. We indulge them with more passive screentime than ever before. The combined effect of video, social media, and Internet "reading" results in generations of ill-equipped young men and women facing a global marketplace that will, in best case, make them suffer and struggle to get up to speed in the employment market, or worst case, crush their hopes and ruin their lives. Prisons and drug rehabs are full of them.

Effect on Creativity

Creativity is measured in many ways. The best comprehensive studies use the of *Torrance Tests of Creative Thinking* (TTCT). Scientific studies over forty years show that there has been a steady decline in children's creativity since 1990, and in related areas since 1984.The current study sampled 272,599 kindergarten through twelfth-grade students and adults.[52] A wide array of creative abilities were tested, including connecting concepts, seeing different angles, as well as imagination and expression:

> "The significant decrease of Strengths scores since 1990 indicates that...children have become less emotionally expressive, less energetic, less talkative and verbally expressive, less humorous, less imaginative, less unconventional, less lively and passionate, less perceptive, less apt to connect seemingly irrelevant things, less synthesizing, and less likely to see things from a different angle."

As communication-through-technology replaces talking person-to-person, our communication becomes more mechanical and impersonal. People are less creative as they grow more dependent on technologies to communicate.

Additionally, the findings show that since 1984:

1) "people of all ages, kindergartners through adults, have been steadily losing their ability to elaborate upon ideas and detailed and reflective thinking;
2) "people are less motivated to be creative; and
3) "creativity is less encouraged by home, school, and society overall."

Given our increasingly screen-centric lifestyles over the past forty years, these findings are not surprising: As everyone in society grows less creative, no one especially notices the decline, or thinks of encouraging a level of creativity that is no longer known.

Today's diminished levels of creativity affect many important areas of our lives. For example, "The results indicate younger children are tending to grow up more narrow-minded, less intellectually curious, and less open to new experiences." People become less accepting of each other, less able to get along, less interested in opposing viewpoints, unable to process multiple sides of an argument, more insulated, unmotivated when it comes to learning in general. Apply this reality to 300,000,000 people over forty years, and today's realities begin to make perfect sense.

> "The results indicate creative thinking is declining over time among Americans of all ages.... The decline is steady and persistent, from 1990 to present, and ranges across the various components tested by the TTCT. The decline begins in young children, which is especially concerning as it stunts abilities which are supposed to mature over a lifetime."[53]

Another scientific study in 2018, which included a random sample of 40,337 2- to 17-year-old children and adolescents in the US, showed that increased screen use led to loss of curiosity.[54] And yet another study by the National Institutes of Health correlates screentime with lower scores on thinking and language tests.[55]

The most precious abilities are the ones we can develop. For example, IQ cannot be increased by working harder. We're stuck with the IQ we received at birth. But creative abilities are wide open for development and improvement by nurturing and practice. Screentime kills the nurturing and removes the practice. The loss is of special significance for our chances to succeed in life. The study found that "creativity was a better predictor of accomplishments than IQ"—so the drop in creativity hits us where it hurts the most—undermining our future potential.

Effect on Productivity

The past fifty years of increasing screentime leaves us in a cycle of shorter attention spans, weaker concentration, less tolerance for difficult work, and less initiative. The result is the collapse of productivity from an increasingly screen-damaged employment pool.

While US workers put in long hours, we are far from the most productive. "Workers in the US put in more hours than nearly everyone but Koreans" But..."Are Americans the world's most productive employees? Not even close, according to recent research" (based on ratio of GDP to hours worked).[56] Over time, employers and employees alike begin to expect the appearance of work (long hours), but not the substance of work (productivity).

The American phenomenon of "presenteeism" is that "employees are expected to be at their desks not because there's so much work to be done, but simply to show their dedication to the company. 'Our workplace cultures reward face time, people who get in early and stay late, people who eat at their desk.'" We value long hours instead of high productivity.[57]

In academia and in business, people have to be imported from places where brain function is not yet as undermined by screen culture. This is why China, for example, is quickly growing stronger than the US on every front where there is competition: economic, industrial, academic, technological, and political—American TV-watching is more than double that of the Chinese.[58]

"Our unexpected findings illustrate the secret ingredients of China's economic success, and a serious threat to America's ability to compete in the global marketplace: discipline."[59] This finding focuses on self-management and relationship-management, which require basic building blocks of mental and emotional development lacking in people who grew up watching TV.[60]

Does Content Matter?

Studies into TV in particular show that watching TV itself is the problem. It doesn't matter what we watch. The effortless video experience itself deteriorates brain health, and the drug-like pleasantness of it masks the damage as it's happening. Even educational programming delays the child's development.[61] "Numerous studies have found that the actual act of watching TV is even more dangerous and potentially damaging to the brain of the developing child than what's on TV."[62] Regardless of what's on, TV is bad for us and our children.

In an example from Dr. Marie Winn's *Plug-In Drug* ("Sesame Street Revisited"): "Noting that their negative findings about Sesame Street's value were often met with skepticism or outright disbelief by parents of preschool children, the authors offered a number of explanations for parents' refusal to be persuaded that Sesame Street, though a delightful entertainment for young children, does not provide them with a particularly valuable learning experience." "The authors also pointed out the favorable press that Sesame Street had received since its inception."[63]

Parental denial of negative findings about a show they like is to be expected—perception of reality for both parents and children alike has been shaped by screen-based values and guidance, which guides us into more screentime.

If a scientific study tells parents that TV is harmful, but a TV publicity campaign tells them that a TV show is good for their kids, parents will reject studies, and believe the TV publicity instead, because it feels better (suggestibility[64]). They deny the science because they like the show. As we saw in the Knowledge section earlier, people increasingly reject science as a rule when it is uncomfortable, or threatens their comfortable sense of "being right."[65]

The false "reality" is being formed in our brains by screens, while our brain cells are made more receptive, so it displaces science and facts with whatever feels right or feels good. In the final blow against good judgment, the lowered brain function of the parents actually promotes gross overestimation of their judgment.[66] No correction needed—everything is fine as it is.

For these reasons, "regardless of the content, television has abetted the creation of an increasingly disempowered people who are not only sedentary and dependent upon professionals to entertain them, but are also complacent and acquiescent to the required advertising necessary to pay for the expensive technologies and equipment."[67]

People are comfortable with the idea that TV is bad only because of bad content. But they remain in denial that the real harm is in the medium itself, and has little to do with the content. Some studies reinforce this misdirection of concern by focusing on content only. They are not proving that the main problem is in the content, they are simply ignoring the effects of screentime, independent of content. "Less attention has been paid to the basic allure of the small screen—the medium, as opposed to the message."[68]

Generation Neutral

We often hear (and have always heard) that things were better in the good old days: "This generation never experienced what we experienced," or "That generation never learned the hard way like we did," etc. Those clichés are red herrings, dead ends, and off topic for the issues being addressed in this book. Screentime damage is generation neutral.

Instead, we are addressing the clinical facts, resulting from hundreds of studies of hundreds of thousands of people over forty years, that if passive screentime is 20,000 hours, we get 20,000 hours less development; if passive screentime is 10,000 hours, we get 10,000 hours less development. It is simply a scientific fact that a person of any generation with 20,000 hours of screentime will suffer a corresponding impairment in mental function, mental health, and emotional development. This impairment happens to have begun in the mid-twentieth century (with TV), and has worsened year after year through today (with video, social media, smartphones, and pervasive screen devices).

Older generations are not immune to screen damage. Even though some older people had the advantage of more natural, prescreen, real-life upbringing, later they let their habits slip into passive-screen dependency.[69]

Screen damage is not generation-specific, and not culture-specific. It is human-specific. Lowered brain function is a clinical fact from passive screentime, regardless of age, generation, or region. Anyone in any generation in any culture will suffer the same impairments, and have done so increasingly since the 1950s.

Active versus Passive

Before screen-based entertainment, everyone's brain was naturally active every moment of every day. Screen culture changed that foundational premise of human activity. Now, as a rule, people who watch a lot of video are more passive even while not watching.[70] "Active" was the normal state of a human brain from the beginning of human history until the mid-twentieth century. Today "Passive" is the normal state of the human brain.

Entertain Me

The late-twentieth century's most influential contribution to humanity has been the concept of "Being Entertained." Entertainment used to be, and should be, the result of active, self-driven, creative effort on the part of people to entertain themselves, as opposed to passively "being entertained" in front of a screen. As a recent *Scientific Learning* article notes, "If we allow children to have poor quality language experiences, substituting entertainment devices for real human language experiences, there will be casualties … If we allow our children to become socially isolated and distracted by a constant barrage of video entertainment options, there will be casualties."[71]

Reversing the childhood damage from "being entertained" requires relearning how to entertain ourselves in ways that require initiative and effort. One of the greatest gifts we can give our children is a screenfree upbringing with guided or self-driven active play and work. Children with such an advantage will thrive and succeed where others suffer and fail.

But in today's screen-surround environment, steering clear of its undermining influences is a challenge. The solution to screen damage is as difficult as it is simple. Screentime itself has reshaped our brains to repel against the difficult.

Real-Life Experience and Play

Amazon.com founder Jeff Bezos credits his success to resourcefulness he learned away from screens, living in the actual world. He spent every summer from age four to sixteen on an isolated farm. There he learned self-reliance, how to build things, how to fix things. "Each time you have a setback, you're using resilience and resourcefulness, and inventing your way out of a box." He passed the same values to his children. Bezos and his wife have "let their kids use sharp knives since they were four and soon had them wielding power tools, because if they hurt themselves, they'd learn. Jeff says his wife's perspective is 'I'd much rather have a kid with nine fingers than a resourceless kid.'"[72]

Nonscreen, real-life playtime is well known to be vital to children's healthy development. NPR's Science correspondent Jon Hamilton brought some of this together, including research from the University of Lethbridge in Alberta, Canada and Washington State University: "The experience of play changes the connections of the neurons at the front end of your brain...those changes in the prefrontal cortex during childhood that help wire up the brain's executive control center, which has a critical role in regulating emotions, making plans and solving problems...The brain builds new circuits in the prefrontal cortex to help it navigate complex social interactions [and we learn] how to interact with others in positive ways...Without play experience, those neurons aren't changed...play is what prepares a young brain for life, love and even schoolwork...the skills associated with play ultimately lead to better grades...But to produce this sort of brain development, children need to engage in plenty of so-called free play. No coaches, no umpires, no rule books...Whether it's rough-and-tumble play or two kids deciding to build a sand castle together, the kids themselves have to negotiate, what are we going to do in this game? What are the rules we are going to follow?"[73]

This crucial emotional development, problem-solving ability, healthy relationships, working with others as an adult, performing well in any occupation, is lost in people who grow up with a lot of TV and video. Essential life skills are disappearing.

The statistics on video-related loss of brain development are overwhelming: the most conservative estimates being 25,000 hours of lost development by age 24, the average probably being closer to 40,000 per 22 years,[74] and perhaps as much as 60,000 hours by age 21.[75]

Based on these facts—men and women in their 20s today barely attain 8- to 12-year-old-levels of mental and emotional maturity based on prescreen standards. The average child today does not develop fundamental life skills, responsibility, caring, fortitude, discipline, intelligence, reasoning, and imagination, among other areas.[76] These facts are so difficult to accept, we subconsciously adopt a collective code of silence on this topic.

Silence does not diminish the disservice we do to our children when we give them a screen device. As Child Psychiatrist Victoria Dunckley points out in *Psychology Today*, that disservice includes

- "Slower language acquisition, and impaired attention"
- "It will eventually cause damage to the nervous system. Mere 'moderation' of screen-time is often not sufficient to interrupt this vicious, self-perpetuating cycle"
- "Children today are functioning at cognitive levels three years younger than same-age peers did thirty years ago"
- "Growing numbers of psychiatric and neurodevelopmental disorders in children"
- "Children who are kept relatively screen-free ... outperform brighter children who've been exposed to ... even 'typical' amounts of screen-time"
- "Screen-free kids (develop) superior brain integration."[77]

Is Addiction Just a Metaphor?

Screen addiction has become the most pervasive brain-chemical addiction today that people know the least about: increasing stress, isolation, loneliness, and depression. On some level, most of us are aware that there is something wrong with our daily use of screen devices, but we keep using them anyway. Worse, we keep allowing our children to overuse them.

As Rutgers University Professor of Media Studies Robert Kubey explains in his article that focuses on TV use, "Television Addiction is no mere metaphor":

> "Psychologists and psychiatrists formally define substance dependence as a disorder characterized by criteria that include spending a great deal of time using the substance; using it more often than one intends; thinking about reducing use or making repeated unsuccessful efforts to reduce use; giving up important social, family or occupational activities to use it; and reporting withdrawal symptoms when one stops using it. *All these criteria can apply to people who watch a lot of television*." Kubey's study goes on to say,

"Viewing begets more viewing" and "Habit-forming drugs work in similar ways."[78]

We do our best to dismiss the screen's harmful consequences, even while feeling guilty about wasting time with TV and smartphones. That's why, for example, we hear common viewer remarks such as: "'If a television is on, I just can't keep my eyes off it,' 'I don't want to watch as much as I do, but I can't help it,' and 'I feel hypnotized when I watch television.'"[79] We often don't enjoy watching TV, but watch it anyway. So we remain trapped in the cycle of decreased life satisfaction and increased anxiety.[80]

TV addiction has been a serious mental-health problem for fifty years. In the early 2000s video games in every household made the screen addiction much worse. Since 2010 smartphones and social media have pushed the mental health crisis to the limit.

Screen interaction triggers brain-chemical rewards (e.g., dopamine hits), while forcing passivity on the viewer. Being more passive makes the viewer more dependent on the screen to suppress anxiety. Being more dependent increases the tendency to remain glued to the device.[81]

While smartphone addiction is the worst of all, it is also true that self-awareness of being addicted is greater among smartphone users. A US Nonprofit *Screen Education* survey found that "65 percent of teenagers wish that they were better able to self-limit the time they spend on their smartphone. While 26 percent wish someone would limit the time they spend on their smartphone for them as they are unable to do so themselves, 37 percent have tried to persuade a friend to reduce the time they spend on their smartphone."[82]

Child psychologist Richard Freed provides an example of what he says is a typical case in his practice, "Casey says she wishes she could put her phone down. But she can't. 'I'll wake up in the morning and go on Facebook just… because," she says. "It's not like I want to or I don't. I just go on it. I'm, like, forced to. I don't know why. I need to. Facebook takes up my whole life.'"[83] This has become the norm, not the exception.

Addiction spiked due to the convenience and portability of smartphones, combined with social media, and more addictive apps since 2010. The BBC investigated social media addiction in 2018 with revealing results.[84]

Silicon Valley insiders told the BBC, "Social media companies are deliberately addicting users to their products for financial gain." A former Silicon Valley employee said, "Behind every screen on your phone, there are generally like literally a thousand engineers that have worked on this thing to try to make it maximally addicting."

A leading technology engineer admitted, "…many designers were driven to create addictive app features by the business models of the big companies that employed them. In order to get the next round of funding, in order to get your stock price up, the amount of time that people spend on your app has to go up … So, when you put that much pressure on that one number, you're going to start trying to invent new ways of getting people to stay hooked."

A former Facebook employee explained, "You have a business model designed to engage you and get you to basically suck as much time out of your life as possible and then selling that attention to advertisers."

The BBC report also noted that, "Last year Facebook's founding president, Sean Parker, said publicly that the company set out to consume as much user time as possible. He claimed it was 'exploiting a vulnerability in human psychology'… 'The inventors', he said, 'understood this consciously and we did it anyway.'"[85]

Similarly in a *Wall Street Journal* article referencing Facebook revealed that "Its own in-depth research shows a significant teen mental-health issue that Facebook plays down in public."[86]

March 2020 internal research refers to Instagram (owned by Meta, the same company that owns Facebook) as "an addictive product [that] can send teens spiraling toward eating disorders, an unhealthy sense of their own bodies and depression. It warns that the Explore page, which serves users photos and videos curated by an algorithm, can send users deep into content that can be harmful." Addiction is an ever-present theme, as teens regularly reported wanting to spend less time on Instagram, but lacked the self-control to do so. "They often feel 'addicted' and know that what they're seeing is bad for their mental health but feel unable to stop themselves."[87]

We know that the addiction hooks of electronic devices are designed into the devices and apps to keep us addicted. Psychologist Adam Alter, author of *Irresistible: The Rise of Addictive Technology and the Business of Keeping Us Hooked*, notes that there are endless user "hooks" embedded in social media, online shopping, and other addictive interfaces. "The list is long—far longer than it's ever been in human history, and we're only just learning the power of these hooks ... Compared to the clunky tech of the 1990s and early 2000s, modern tech is efficient and addictive."[88]

Instead of addressing app addiction, every year sees more addictive apps gaining popularity, such as TikTok and BeReal, two of the most popular apps downloaded in 2022.[89]

A Forbes article "Digital Crack Cocaine: The Science Behind TikTok's Success" describes the escalation of the same addictiveness we've seen in other social apps: "platforms like TikTok—including Instagram, Snapchat and Facebook—have adopted the same principles that have made gambling addictive." In the article, USC professor and author Dr. Julie Albright describes the experience: "'You'll just be in this pleasurable dopamine state, carried away. It's almost hypnotic, you'll keep watching and watching.'" Albright notes that "'Our brains are changing based on this interaction with digital technologies and one of these is time compression ... Our attention spans are lowering.'"[90]

The gambling comparison is worth mention. As more states legalize sports betting in the 2020s, and betting apps proliferate on our phones, we have all new opportunities for screen-related addiction. Americans placed an average of under $1 billion per month in bets during the first half of 2019. By 2022 in that same period, betting had increased to nearly $8 billion per month.[91]

New York Times investigative reporter Rebecca R. Ruiz reported in a December 2022 article that she herself had downloaded three betting apps and soon had placed over 100 bets across the three platforms: "One in five Americans bet on sports this year. Against most odds—my total lack of interest in all professional athletics—I was one of them. And I was hooked."[92]

Content on screens can lead to poor self-image, depression, and other mental health issues. But screen addiction itself is content-neutral.

Our brains are no match for the algorithms, hooks, addictive design, and other engineered mechanisms that manipulate our brain chemicals and control our attention. We simply learn to live with emotional exhaustion, depression, feeling out of control, and persistent stress and anxiety.[93]

The organization Swipe Left For Addiction led by Psychologist and Behavioral Scientist Marc Atherton notes that addictive design is related to the Skinner Box. Lab rats push the levers compulsively when the rats are sometimes rewarded, and sometimes not rewarded. The unpredictability of reward increases the dopamine hits to the brain, which drives the addictive behavior. This is what happens when we check social media for likes, comments, mentions, or check for email, or any other onscreen reinforcement. We are pressing the lever for a dopamine hit. "The more we check, the more the behaviour is reinforced and the compulsive loop is created."

It reduces our cognitive function to that of a laboratory mouse or rat, with no thought of resisting, no thought at all. "Product designers are programming their apps to perfectly time rewards so they can grab as much of our time as possible as they compete in the Attention Economy."[94]

The brightest minds in technology today are not helping people deal with technology better—quite the opposite—they are doing everything they can to make the screen takeover our lives, to pull us into irresistible addiction: "According to Tristan Harris, a 'design ethicist,' the problem isn't that people lack willpower; it's that 'there are a thousand people on the other side of the screen whose job it is to break down the self-regulation you have.'"[95]

The intended addictiveness is so effective, experts refer to it as being "weaponized" for addiction: "The people who create and refine tech, games, and interactive experiences are very good at what they do. They run thousands of tests with millions of users to learn which tweaks work and which ones don't—which background colors, fonts, and audio tones maximize engagement and minimize frustration. As an experience evolves, it becomes an irresistible, *weaponized* version of the experience it once was."[96]

Addictive behaviors have become so common they are now mainstream. "These new addictions don't involve the ingestion of a substance … but they produce the same effects because they're compelling and well designed … they've all become progressively more difficult to resist."[97] Almost everyone we meet is losing a large part of their life to a screen-related addiction.

Our addiction, and the developmental deterioration that goes with it, is so pervasive that we stopped noticing. Besides hurting ourselves, we also hurt our children. Being screen-dependent as parents, we accept the mental and physical damage to our children, even while we know at some level that it's harmful. Placing screens in front of our children helps us "normalize" the behavior so we feel better about ourselves. We soothe our conscience while our children pay the mental-health price—we perpetuate the damage and ruin their potential.

Even by 2008 when smartphones were relatively new, the addictive effects were being noticed. According to a British study commissioned by *Royal Mail* and carried out by *YouGov* in 2008, 53 percent of Brits felt anxious when they "lose their mobile phone, run out of battery or credit, or have no network coverage." Ten years later in 2018, multiple studies have proved that this number is significantly higher. "And with children growing up today who haven't had any other experience but 24/7 access to the online world…it's easy to see how the number of people suffering from separation anxiety when away from their phones will only go up."[98]

A 2015 Pew Research Center study found compelling evidence of addiction on the rise: "44% of smartphone owners had a problem doing something they needed to do when they didn't have their phone with them…46% of Americans said they believed that they just 'couldn't live without their phones.'"[99]

Addiction and its mental-health problems are only getting worse. The new condition specific to smartphones is called "nomophobia," which literally means no-mobile-phone-phobia: "People that suffer from an overwhelming, crippling, anxiety when their phone has died or has no signal."[100]

Dr. Brenda K. Wiederhold of the Interactive Media Institute, San Diego, and the Virtual Reality Medical Institute, Brussels, describes nomophobia this way in a 2017 study: "'fear of missing out, and fear of being offline—all anxieties born of our new high-tech lifestyles—may be treated similarly to other more traditional phobias. Exposure therapy, in this case turning off technology periodically, can teach individuals to reduce anxiety and become comfortable with periods of disconnectedness.'"[101]

Screen addiction goes deeper. Dr. Nicholas Kardaras, clinical psychologist and specialist in neurophysiology and the treatment of addiction, explains:

> "Over 200 peer-reviewed studies correlate excessive screen usage with a whole host of clinical disorders, including addiction. Recent brain-imaging research confirms that glowing screens affect the brain's frontal cortex — which controls executive functioning, including impulse control — in exactly the same way that drugs like cocaine and heroin do. Thanks to research from the US military, we also know that screens and video games can literally affect the brain like digital morphine."[102]

Exceptional people are no exception. Recent *Washington Post* research cited by Dr. Jim Taylor, an internationally recognized authority on the psychology of sport and parenting, reported that professional basketball players suffer device addiction like the rest of us. "The best basketball players in the world are getting beaten by their phones. They check their phones during team meetings, before games, at half time, and after games. *They know they should stop, but they can't.* Now, apply those same social-media habits to your young athletes and ask yourself whether you think it might be time to help them develop a healthier relationship with their technology."[103]

Professional athletes who provide role models are falling into the same deterioration of skill, discipline, and will, from the same device addiction as everyone else.

Addiction to screens is not a metaphor. Screen addiction is a brain-chemical addiction like any drug addiction. Our obsessive screen usage has damaged brain function, mental health, emotional development and physical health of multiple generations today.

The Tech Industry's War on Kids

Child and adolescent psychologist Richard Freed provided an extensive report on the seriousness of screen-related damage to children's brains, and most disturbingly, how intentional it is. It is a long article with a wealth of information. Reading the full article is recommended[104], but here are a few key excerpts:

- "The parents I work with simply have no idea about the immense amount of financial and psychological firepower aimed at their children to keep them playing video games 'forever.'"
- "Nestled in an unremarkable building on the Stanford University campus in Palo Alto, California, is the Stanford Persuasive Technology Lab, founded in 1998. The lab's creator, Dr. B.J. Fogg, is a psychologist and the father of persuasive technology, a discipline in which digital machines and apps—including smartphones, social media, and video games—are configured to alter human thoughts and behaviors. As the lab's website boldly proclaims: 'Machines designed to change humans.'"
- "In the Venice region of Los Angeles, now dubbed 'Silicon Beach,' the startup Dopamine Labs boasts about its use of persuasive techniques to increase profits: 'Connect your app to our Persuasive AI [Artificial Intelligence] and lift your engagement and revenue up to 30% by giving your users our perfect bursts of dopamine,' and 'A burst of Dopamine doesn't just feel good: it's proven to re-wire user behavior and habits.'"
- "Ramsay Brown, the founder of Dopamine Labs, says in a KQED Science article, 'We have now developed a rigorous technology of the human mind, and that is both exciting and terrifying. We have the ability to twiddle some knobs in a machine learning dashboard we build, and around the world hundreds of thousands of people are going to quietly change their behavior in ways that, unbeknownst to them, feel second-nature but are really by design.' Programmers call this 'brain hacking,' as it compels users to spend more time on sites even though they mistakenly believe it's strictly due to their own conscious choices."

- "Banks of computers employ AI to 'learn' which of a countless number of persuasive design elements will keep users hooked. A persuasion profile of a particular user's unique vulnerabilities is developed in real time and exploited to keep users on the site and make them return again and again for longer periods of time. This drives up profits for consumer internet companies whose revenue is based on how much their products are used."
- "Persuasive technology's use of digital media to target children, deploying the weapon of psychological manipulation at just the right moment, is what makes it so powerful. These design techniques provide tech corporations a window into kids' hearts and minds to measure their particular vulnerabilities, which can then be used to control their behavior as consumers."
- "Persuasive technologies are reshaping childhood, luring kids away from family and schoolwork to spend more and more of their lives sitting before screens and phones. According to a Kaiser Family Foundation report, younger U.S. children now spend 5 ½ hours each day with entertainment technologies, including video games, social media, and online videos. Even more, the average teen now spends an incredible 8 hours each day playing with screens and phones. Productive uses of technology—where persuasive design is much less a factor—are almost an afterthought, as U.S. kids only spend *16 minutes* each day using the computer at home for school."
- "With each passing day, new and more influential persuasive technologies are being deployed."[105]

A Chilling Episode

Fifty years ago, TV addiction should have taught us lessons about the dangers of passive screen entertainment, long before the Internet, smartphones and tablets. Many studies, some of which are cited in this book, revealed TV's damage to brain function, mental health, and emotional development long ago. Those original studies were warnings, which we ignored.

Instead of responding constructively by limiting screentime, we embraced screen use so that it became pervasive. We gradually became addicted to screen entertainment, and its pervasiveness gradually normalized screen addiction. Screen addiction is now normal everyday behavior.

The below episode illustrates one of those early warnings that we ignored, about how insidious screentime had become, especially for children:

> In the tragic 1970s example from Dr. Marie Winn's "A Chilling Episode" (*Plug-in Drug*), some parents chose to drug their children to treat screen addiction symptoms, rather than cure the disease by reducing screentime. In this particular case in point, many children were suffering "Tired-Child Syndrome"—"chronic fatigue, loss of appetite, headache, and vomiting"—doctors learned the children were "spending three to six hours of watching television daily, and six to ten hours on weekends." The doctors immediately prescribed removal of TV.
>
> "The effects were dramatic for the 12 children whose parents followed the instructions fully: the symptoms vanished within two to three weeks."[106] It goes on to say the kids whose parents reduced TV but did not eliminate it, achieved proportional partial recovery, or recovery over longer period of time.
>
> The tragic part of the story is that many parents ultimately surrendered their children to screen addiction at the expense of their children's mental health. Even after witnessing the amazing recovery and renewed health of their children, they later let their children go back to their old TV habits. Their children's symptoms returned. This time, however, the parents put their children on chlorpromazine (antipsychotic medication) to treat the symptoms rather than limit TV.[107] Surrender to screen addiction was so complete, the parents condemned their own children to a drugged state of helpless brain deterioration, stunted mental and physical development, and all of this with full knowledge of the damage they were doing to their children. They did it because it was easier than resisting screen addiction.

Because of a today's almost total screen-surround lifestyle, we experience more brain reshaping than even the previous fifty years:

"Social media has *completely shaped the brains* of the younger people I work with … the result is a landscape filled with disconnection and addiction."[108]

What Technology Leaders Say

In technology circles, the extreme addictiveness of electronic devices, apps, and programs is well known. That's why technology leaders are the ones who most restrict screentime of their own children. They know that the addictiveness is irresistible, and that the addiction will undermine their children's brain capabilities and mental health. Senior Editor Chris Weller cites examples in a *Business Insider* article, "An MIT psychologist explains why so many tech moguls send their kids to anti-tech schools."[109] Professor of the Social Studies of Science and Technology at the Massachusetts Institute of Technology (MIT), Dr. Sherry Turkle—author of numerous books on the negative social effects of technology—notes that "Technology moguls like Bill Gates, Steve Jobs, and other high-powered entrepreneurs tend to share similar qualities: persistence, ingenuity, grit, just to name a few. But one of the more surprising traits is the philosophy that kids ought to be raised tech-free."[110]

There are plenty of examples of tech leaders keeping their children screenfree. New York University psychology professor Adam Alter tells a revealing story in a recent Ted Talk. Alter relates how Steve Jobs had said the "iPad is the best browsing experience you've ever had." In an interview, with Jobs, a journalist said, "Your kids must love the iPad." Jobs replied, "They haven't used it."[111] Steve Jobs was one of a large number of technology leaders, including others at Apple, eBay, Google, Yahoo, and Hewlett-Packard, who have kept their children away from screens.[112]

Here are some examples Alter provides of tech leaders' attitudes towards kids and screens:

- "[Steve Jobs] kept the iPad from his kids because, for all the advantages that made them unlikely substance addicts, he knew they were susceptible to the iPad's charms.

- "Evan Williams, a founder of Blogger, Twitter, and Medium, bought hundreds of books for his two young sons, but refused to give them an iPad.
- "Lesley Gold, the founder of an analytics company, imposed a strict no-screen-time-during-the-week rule on her kids.
- "Chris Anderson, the former editor of WIRED, enforced strict time limits on every device in his home, 'because we have seen the dangers of technology firsthand.' His five children were never allowed to use screens in their bedrooms."

Screen devices are sold to us as a beneficial technology. But technology leaders quietly keep their own children away from screens. "These entrepreneurs recognize that the tools they promote—engineered to be irresistible—will ensnare users indiscriminately."[113]

Alter also points out that it's no accident the Waldorf School in Silicon Valley doesn't introduce screens to kids at all until 8th grade, and 75 percent of the Waldorf students' parents are high-level silicon valley tech executives.[114]

According to a *New York Times* report on a Silicon Valley Waldorf School, "the school's chief teaching tools are anything but high-tech: pens and paper, knitting needles and, occasionally, mud."

It is very telling that this is the environment technology leaders demand for their children:

- "Not a computer to be found. No screens at all. They are not allowed in the classroom, and the school even frowns on their use at home." The schools philosophy is that "computers inhibit creative thinking, movement, human interaction and attention spans."
- "Schools nationwide have rushed to supply their classrooms with computers, and many policy makers say it is foolish to do otherwise. But … at the epicenter of the tech economy, parents and educators have a message: computers and schools don't mix….While other schools in the region brag about their wired classrooms, the Waldorf school embraces a simple, retro look—blackboards with colorful chalk, bookshelves with encyclopedias, wooden desks filled with workbooks and No. 2 pencils." The proof is in the pudding, and 94 percent of the students go on to college. "Parents of students at the

Los Altos school say it attracts great teachers who go through extensive training in the Waldorf approach, creating a strong sense of mission that can be lacking in other schools."

- "Paul Thomas, a former teacher and an associate professor of education at Furman University, who has written 12 books about public educational methods (says), 'a spare approach to technology in the classroom will always benefit learning.' … 'Teaching is a human experience,' he said. 'Technology is a distraction when we need literacy, numeracy and critical thinking.'"
- Where advocates for stocking classrooms with technology say children need computer time to compete in the modern world, Waldorf parents, such as Alan Eagle, a Waldorf parent who works in executive communications at Google, counter: What's the rush, given how easy it is to pick up those skills? "It's supereasy. It's like learning to use toothpaste" … "At Google and all these places, we make technology as brain-dead easy to use as possible. There's no reason why kids can't figure it out when they get older."[115]

If you want to know what's best for your children in terms of technology, look at how the technology experts educate their own children.

The Rich Get Smart, The Poor Get Technology

This section's title is taken from an article of the same title, "The Rich Get Smart, The Poor Get Technology: The New Digital Divide in School Choice,"[116] The article highlights that fact that feel-good technology solutions for underprivileged students is a cheap way to make it look like we're addressing educational inequality. In fact, screens in the classroom will only widen the educational divide between rich and poor.

As *The New York Times* reports, "It wasn't long ago that the worry was that rich students would have access to the internet earlier, gaining tech skills and creating a digital divide." But the opposite has become the reality: "Children of poorer and middle-class parents will be raised by screens, while the children of Silicon Valley's elite will be going back to wooden toys and the luxury of human interaction."[117]

The new digital divide has already played out and the inequality grows wider every day. For example, "Throwback play-based preschools are trending in affluent neighborhoods," while poorer kids get state-funded online-only preschool and more federal grants to expand screen-based schools. There is already a growing screentime differentiator: middle-class and poorer children watch screens significantly more than their wealthier counterparts.

And according to early results of a landmark study on brain development of more than 11,000 children that the National Institutes of Health is supporting, "children who spent more than two hours a day looking at a screen got lower scores on thinking and language tests."[118] Psychologist Richard Freed, who wrote a book about the dangers of screentime for children, warns us that screen devices are "too relied upon in schools for low-income children. And he sees the divide every day as he meets tech-addicted children of middle and low-income families."[119]

Vocabulary correlates to thinking ability, and low-income kids have less vocabulary, because low-income kids engage in significantly more screentime than more affluent kids. The gap is alarming and widening.

As NPR Senior Editor Cory Turner points out in a recent article, "...by the end of age 3, children from low-socioeconomic backgrounds will have heard 30 million fewer words than their more affluent peers. And this number itself was correlated not just with differences in vocabulary but also differences in IQ and test scores..." This translates to a severe impact on brain development and future success. "In the first three years of life, you'll have no more rapid and robust brain growth than during that time. It's when 80 to 85 percent of the physical brain develops."[120]

Perhaps the new digital divide should not be surprising. Even after controlling for confounding factors such as socio-economic, we know that as TV in childhood increases, the chances of dropping out of school increase; and chances of getting a college degree decrease.[121] That's why more affluent families are banning screentime for their kids. They are responding to the growing body of evidence showing the harmfulness of screentime.

More affluent families are in a better position to protect their children's development, with advantages such as affording private screenfree schools, access to expensive nonscreen recreational options for their kids, having more time for in-person interaction with kids as opposed to single moms who rely on screens to occupy their children because they work two jobs, for example.

Specifically addressing inequality in the classroom, an English study published in *Labour Economics* found that banning phone devices from the classroom "can be a low-cost policy to reduce educational inequalities."[122]

Dr. Nicholas Kardaras, clinical psychologist and specialist in neurophysiology and the treatment of addiction, notes that there is a mountain of evidence now showing that screen devices in the classroom have a negative impact on learning. This evidence comes from forty-eight studies, one in 2015 alone included 140,000 students.[123]

The misguided panacea of throwing technology at disadvantaged students feels good but doesn't address educational inequality. It only widens the gap between the rich, who get human interaction and real-world learning, versus the poor, who get screen learning, which impairs development and causes lower test scores. When you see socio-economic high-end schools doing one thing and getting excellent results, and the economically disadvantaged doing a different thing and getting terrible results, we should pay attention to that "thing" whatever it is: In this case, the use/nonuse of screens.

The pattern keeps repeating itself: Each new study, expecting to discover benefits of technology in the classroom, discover the opposite is true: *The Rich Get Smart, The Poor Get Technology!*[124]

Screentime and the Decline of Mental Health

Depression and loneliness are two symptoms most notably associated with smartphone usage. Even TV didn't replace in-person relations. It reduced them, but didn't replace them. People couldn't take a TV with them everywhere. Smartphones and tablets let us take TV, movies, games, and social media everywhere we go. With our human-less screen-surround lifestyle, whether in a cabin in the woods or in a high-rise office in New York City, the life experience is the same, in that there is no life experience happening in either location. Screens have become the universal experience for everyone everywhere.

In-person social interaction is critical for mental health, but screentime today almost totally displaces in-person time. It's not just a slight tendency to feeling alone. As San Diego State University professor and psychologist Jean Twenge explains, "For some people it might be tempting to dismiss this as teen girls being sad, but we're looking at clinical-level depression that requires treatment. We're talking about self-harm that lands people in the ER."[125]

Time spent with screen devices directly links to suicidal thoughts and clinical depression. As screentime increases, the rise in depression steadily grows.

According to Dr. Twenge, "the effect of screen activities is unmistakable: The more time teens spend looking at screens, the more likely they are to report symptoms of depression." In studies lasting from 1975 to 2018, "There's not a single exception. All screen activities are linked to less happiness, and all nonscreen activities are linked to more happiness."[126]

Dr. Twenge saw a "spike in teen mental health issues between 2011 and 2015. The spike, she explains, was sudden, with major depressive episodes among teens increasing by 50 percent within those few years." The timing corresponds to the introduction and popularization of smartphones. Several studies mirror the correlation of smartphone use with major depression, self-harm, and suicide.[127] In a 2018 study — "In nationally representative yearly surveys of United States 8th, 10th, and 12th graders 1991–2016 (N = 1.1 million), psychological well-being (measured by self-esteem, life satisfaction, and happiness) suddenly decreased after 2012 [as smartphones were spreading]. Adolescents who spent more time on electronic communication and screens (e.g., social media, the Internet, texting, gaming) and less time on nonscreen activities (e.g., in-person social interaction, sports/exercise, homework, attending religious services) had lower psychological well-being."[128]

Twenge found that "Teens who spend three hours a day or more on electronic devices are 35 percent more likely to have a risk factor for suicide, such as making a suicide plan," and adds that "three times as many 12-to-14-year-old girls killed themselves in 2015 as in 2007, compared with twice as many boys."[129]

Also noted in a *Clinical Psychological Science* report, mental health deteriorated rapidly across a generation starting in 2010 to 2011 when they started using smartphones. "Since 2010, adolescents spent more time on social media and electronic devices, activities positively correlated with depressive symptoms and suicide-related outcomes. Over the same years, adolescents spent less time on nonscreen activities such as in-person social interaction, print media, sports/exercise, and attending religious services, activities negatively correlated with depressive symptoms [and] iGen adolescents in the 2010s spent more time on electronic communication and less time on in-person interaction than their Millennial and Generation X (GenX) predecessors at the same age."[130] (See the endnote for a table showing birth-year ranges for seven generations.[131])

As executive director of UC Berkeley's *Greater Good Science Center* Christine Carter notes, the best predictor of a person's happiness and well-being is having a large breadth and depth of real-life social connections. "Not social *media* connections, not a large network of *online* friends and family—those things do not predict happiness. But feeling like we are a part of something larger than ourselves, feeling deeply embedded in a community of friends and family whom we see face-to-face."[132]

Carter confirms research that shows, as she describes it, "the blazingly obvious, that the more time kids spend online and on their devices, the less time they spend interacting in face-to-face encounters. Screen time is a zero-sum game: Social media, video games, online browsing, and other new media take the place of playing sports, going to parties, or just hanging out with their friends."[133]

Carter reinforces Dr. Twenge's findings as well, "So today's teenagers feel lonely because they actually *are* alone more. 'The number of teens who get together with their friends every day has been cut in half in just 15 years, with especially steep declines recently,' writes research psychologist and demographer Jean Twenge."[134]

The correlation between screentime and mental health decline was well documented by 2012, as Victoria L. Dunckley M.D., explains in her article on modern-day screentime disorders. Screentime disorders typically include the following symptoms:

> "The child exhibits symptoms related to mood, anxiety, cognition, behavior, or social interactions that cause significant impairment in school, at home, or with peers. Typical signs/symptoms mimic chronic stress and include irritable, depressed or labile mood, excessive tantrums, low frustration tolerance, poor self-regulation, disorganized behavior, oppositional-defiant behaviors, poor sportsmanship, social immaturity, poor eye contact, insomnia/non-restorative sleep, learning difficulties, and poor short-term memory."

Dunckley notes that the symptoms can be reversed:

> "Symptoms markedly improve or resolve with strict removal of electronic media (an 'electronic fast'); three- to four-week electronic fasts are often sufficient but longer fasts may be required in severe cases."

Dunckley also warns against relapses:

> "Symptoms may return with re-introduction of electronic media following a fast, depending on a variety of factors. Some children can tolerate moderation after a fast, while others seem to relapse immediately if re-exposed."[135]

Educators are on the frontlines, with classrooms full of students every day. What do they see? "In a nationally representative survey by the EdWeek Research Center, 88 percent of educators reported that in their experience, students' learning challenges rose along with their increased screentime. Moreover, 80 percent of educators said student behavior worsened with more screentime. Over a third said student behavior has gotten 'much worse' due to rising screentime."[136]

Yet another scientific study in 2018, which included a random sample of 40,337 2- to 17-year-old children and adolescents in the US, showed that increased screen use led to elevated anxiety, depression, and emotional instability.[137]

Dr. Twenge warns, "It's not an exaggeration to describe iGen as being on the brink of the worst mental-health crisis in decades. Much of this deterioration can be traced to their phones."[138]

It is no surprise that iGen in particular are less able to manage stress, fear, and trepidation. These cause exaggerated hesitancy, which keeps them from performing as well as they could.[139]

As noted earlier in this section, screentime causes loneliness—and loneliness increases stress above and beyond the direct screentime-stress correlation. "This hardly needs to be said, but loneliness and social isolation cause profound distress. Feeling lonely causes your cortisol, a stress hormone, to skyrocket. Experiments by neuroscientist John Cacioppo have shown that feeling acutely lonely is as stressful as experiencing a physical attack."[140]

Another recent study in clinical psychology reinforces the findings that "adolescents who spent more time on new media (including social media and smartphones) were more likely to report mental health issues, and adolescents who spent more time on nonscreen activities were less likely."[141]

Being well established scientifically, the social-media related mental-health crisis is now being recognized in legal quarters as well. The case of a tragic 2017 suicide of 14-year-old Molly Russell was heard in a UK court in September 2022. It was characterized as a David and Goliath struggle—"a legal battle that pitted the Russell family against some of Silicon Valley's largest companies." In the hearing: "Sitting in the witness box of a small London courtroom this week, a Meta executive faced an uncomfortable question: Did her company contribute to the suicide of a 14-year-old named Molly Russell?" The court concluded, "'Molly Rose Russell died from an act of self-harm while suffering from depression and the negative effects of online content,'" … Rather than officially classify her death a suicide, he said the internet 'affected her mental health in a negative way and contributed to her death in a more than minimal way.'"[142]

Similar findings come from research reported in "Facebook Use Predicts Declines in Subjective Well-Being in Young Adults."[143]

Clinical psychologist and behavioral therapist Dr. Gadi Lissak assembled a report from dozens of studies conducted at research centers around the world. He describes the physical, physiological, psychological, sociological, and neurological adverse consequences from screentime on children. The studies found that screentime causes an increase in depressive and suicidal symptoms, antisocial behavior and decreased prosocial behavior, ADHD-related behavior, brain-structural changes related to cognitive control and emotional regulation, craving behavior which resembles substance dependence behavior, and decreased social coping.[144]

A clinical study by University of Pennsylvania psychologist Melissa Hunt, reported in the *Journal of Social and Clinical Psychology*, reinforces the growing abundance of research in this area. The study also found that as social media use increases, so does loneliness and depression. The study also found that in just three weeks, discontinued or limited use of social media led to "significant reductions in loneliness and depression."[145]

Social media may be useful for reaching faraway friends and staying in touch with relatives. But it is a terrible substitute for in-person relationships. As the Twenge study in *Clinical Psychological Science* notes, "In-person social interaction provides more emotional closeness than electronic communication and … is more protective against loneliness," whereas, electronic communication, particularly social media, increase feelings of loneliness.[146]

Our isolation is pronounced especially when we are in a crowd or in a social setting "around people" but still not "with people," because of our device addiction. The phenomenon is powerfully depicted in Eric Pickersgill's photo series of people missing out on life, missing out on the joys of being with other people, losing the essence of life-experience that once made us human.[147]

As child and adolescent psychologist Richard Freed observes, "as the typical age when kids get their first smartphone has fallen to 10, it's no surprise to see serious psychiatric problems—once the domain of teens—now enveloping young kids. Self-inflicted injuries, such as cutting, that are serious enough to require treatment in an emergency room, have increased dramatically in 10- to 14-year-old girls, up 19% per year since 2009."[148]

There is no longer any question that screen devices and social media lead to depression and many other mental health problems.[149] As in this example from a *BBC* study, the results are, "I'm feeling lonely, 'Let me check my phone.' I'm feeling insecure, 'Let me check my phone.'"[150] We suffer the consequences of lost time, lost productivity, lost skills, lost relationships, lost happiness. The steady diet of screentime steadily deteriorates brain function.[151] As a remedy for screen-related loneliness, we turn to the screen to deliver temporary relief in the form of dopamine hits—which ultimately makes us more lonely and depressed.

As described in an *NPR* piece "Facebook Makes Us Sadder And Less Satisfied, Study Finds":

> "If you're feeling bummed, researchers did test for and find a solution. The prescription for Facebook despair is less Facebook. Researchers found that face-to-face or phone interaction … had the opposite effect. Direct interactions with other human beings led people to feel better."[152]

Talking to Strangers

The desperation of clinical depression, or just plain loneliness, is a tragic reality that's more prevalent today than ever before, because of social media and smartphones.[153] More in-person interaction could improve our mental health, promote a sense of closeness to each other, and reduce loneliness.[154] In the past, this closeness was not just with family and friends, it included talking to strangers as part of daily life.

Before screens isolated us, people felt much more comfortable talking to strangers in public places. It was normal, natural, expected behavior throughout history. Comfortable conversation with strangers gradually became less natural as we became more isolated during the TV generations of the mid-to-late-twentieth century. That loss has worsened in recent decades, and most severely since 2010. Because of frequent, extensive interaction with strangers as well as acquaintances in the past, isolation-related depression didn't happen as much. "In the past, you may go out and meet with your friends and talk about something, but when you got home you'd go to sleep."[155] Today, people might go out to a coffee shop but keep their heads down staring at their laptop or smartphone, then go home and spend the evening on social media. Human interaction is not part of life.

As the habit of talking to strangers faded from our daily life, so our expectations faded. As *New York Times* columnist David Brooks explains in an August 2022 column reviewing several related studies, he notes that we underestimate how much we would enjoy talking to strangers.[156] Hence, we don't do it as much, and we suffer for it, as our reluctance undermines our own well-being. In a survey conducted by University of Chicago Professor of Behavior Science Nicholas Epley, "only 7 percent of people said they would talk to a stranger in a waiting room. Only 24 percent said they would talk to a stranger on a train."[157] Dr. Epley's behavioral study on the subject found that, "Commuters expected to have less pleasant rides if they tried to strike up a conversation with a stranger. But their actual experience was precisely the opposite. People randomly assigned to talk with a stranger enjoyed their trips consistently more than those instructed to keep to themselves. Introverts sometimes go into these situations with particularly low expectations, but both introverts and extroverts tended to enjoy conversations more than riding solo."[158]

Additional studies expand on these findings. For example, a National Academy of Sciences study shows that "most people underestimate how much they will learn from conversations with strangers." Behavioral Sciences Professor Stav Atir, with others, conducted seven experiments, which showed that "people may systematically underestimate the informational benefit of conversation, creating a barrier to talking with—and hence learning from—others in daily life."[159]

It doesn't end there. A *Journal of Personality and Social Psychology* study revealed that we underestimate how much we would enjoy longer conversations with new acquaintances. "Participants expected their enjoyment to decline as their conversations continued, but experienced stable or increasing enjoyment in reality. This miscalibration arose at least partly because participants underestimated how much they would have to discuss." Leaving this mistaken assumption uncorrected undermined well-being.[160] We try to avoid an unpleasant experience—mistakenly by avoiding longer conversations with a stranger—the very thing that will most benefit our life experience and mental health.

In addition to dreading longer conversations, we dread deeper conversations as well. Our misconceptions about the former, also apply to the latter. A separate study published in the *Journal of Personality and Social Psychology* found that "people underestimated how much they're going to enjoy deeper conversations compared to shallower conversations." The study concluded, "People may want deep and meaningful relationships with others, but may also be reluctant to engage in the deep and meaningful conversations with strangers that could create those relationships."[161]

A conversation may not cure clinical depression, but the recent spike in loneliness and depression is a direct result of the decline in human interaction. That decline is a direct result of the screentime that takes up most of our waking hours. When we feel depressed or lonely, we turn to the very devices that further isolate us, which deepens our depression and loneliness.

Physical Contact

Whether talking to strangers, or gathering with friends, physical interaction is another part of the once-common human contact that we have lost.

As *Emotional Intelligence 2.0* author and clinical psychologist Travis Bradberry notes, "When you touch someone during a conversation, you release oxytocin in their brain, a neurotransmitter that makes their brain associate you with trust and a slew of other positive feelings. A simple touch on the shoulder, a hug, or a friendly handshake is all it takes to release oxytocin. … Just remember, relationships are built not just from words, but also from general feelings about each other. Touching someone appropriately is a great way to show you care."[162]

A *Clinical Psychological Science* study echoes similar findings: "It is worth remembering that humans' neural architecture evolved under conditions of close, mostly continuous face-to-face contact with others … including touch … and that a decrease in or removal of a system's key inputs may risk destabilization of the system."[163]

Our isolation means we're no longer familiar with daily or hourly physical contact. Lost familiarity makes us afraid to touch each other in the healthy human ways that people did for thousands of years until the late-twentieth century and, as noted in earlier sections, especially since 2010. We are more awkward in any setting of face-to-face interaction. Almost nothing about human interaction comes naturally anymore. Everyone is self-conscious and afraid that everything about human contact is inappropriate because everything about human interaction has become unfamiliar, awkward, and strange.

What have we lost? Feeling comfortable striking up a conversation with a stranger, the automatic sense of natural camaraderie with others, and daily friendly physical contact, the feeling that we're all in this together, and overall brain health.[164] We have lost a lot. This "destabilization of the system" refers to ourselves, our communities, and our society. As noted in the "Talking to Strangers" section, when we feel depressed or lonely, we turn to the very devices that further isolate us, which deepens our depression and loneliness.

Company Loves Misery

In addition to the many facets of mental illness we have seen so far, social media gives us a desire to see misfortune in others. People today feel better seeing other people's problems, and feel worse seeing other people's happiness.[165] Is this new prevailing attitude surprising? Consider the factors:

- Video "reality" reduces real-time with friends and reduces exposure to friends' problems.
- Video since childhood leaves us more susceptible to accept video messages as true.[166]
- Social media emphasizes how much more interesting and exciting other people are.
- Isolation prevents corrective, balanced, realistic comparison.
- Social media time correlates to depression and suicidal thoughts.[167]

When someone posts how they failed at something, people soak it in like nourishment. What a relief to see that someone else has a problem!

The Anxiety Effect

Frequent TV-and-video viewers have worse anxiety as a normal part of their life when away from the screen. Non-viewers or light viewers are happier, and do not experience that anxiety when a screen is not available. Regular viewers require external stimulation at all times. As Robert Kubey notes in his *Scientific American* article, "We wondered whether heavy viewers might experience life differently than light viewers do. ... What we found nearly leaped off the page at us. Heavy viewers report feeling significantly more anxious and less happy than light viewers do in unstructured situations...."[168]

As screen use increases, anxiety increases, until we become helpless without a screen device to entertain us. As described by Dr. Nicholas Kardaras, clinical psychologist and specialist in neurophysiology and the treatment of addiction, "We see the aggressive temper tantrums when the devices are taken away and the wandering attention spans when children are not perpetually stimulated by their hyper-arousing devices. Worse, we see children who become bored, apathetic, uninteresting and uninterested when not plugged in." This is just the latest in a long history of evidence. "In addition, hundreds of clinical studies show that screens increase depression, anxiety and aggression and can even lead to psychotic-like features…."[169]

When viewers separate from external screen stimulation, and then go into the real world, frustrations become more prevalent, because they no longer possess the ability to cope with life without screen-fed stimulation. Viewers can no longer generate their own brain activity internally. Many people's response to this frustration is to restore temporary brain-comfort with more passive screentime. It becomes a conditioned response to escape the overwhelming anxieties and irritations that arise from normal activity. While it provides an escape, it also further erodes our ability to cope with problems. Life becomes effortless under the calm induced by the screen, temporarily. During this passive relaxation, our brain atrophies further.[170]

The Stress Effect

People who watch a lot of TV feel more stress.[171] Having to talk to someone is more stressful, being alone is more stressful. There is no cause for the extra stress, other than a reduced capacity to manage stress. During passive screentime, we are not developing stress-management areas of the brain, like we do when we are more actively engaged in life.

Even when there is nothing stressful in a person's life, TV causes an increase in stress levels. Similar to the anxiety conditioned response, people seek the comfort of the screen to alleviate stress, not realizing it is the cause. Once again, the act of watching TV increases the need to watch more.[172]

Stress is often associated with having a busy schedule. Screen-related stress creates the illusion of being busy. We claim to be too busy, that life is too hectic, when in fact we are not too busy.

"Most of us are a lot less busy than we claim we are…in general we have 30–40 hours of free time each week…It's very popular, the feeling that there are too many things going on...But the evidence does not back it up."[173]

By spending 30–40 hours a week immersed in passive screentime, the stress becomes real. By wasting all that time, suddenly we are behind in everything and being busy becomes real.

"No one likes to think of themselves as self-deluded, but the 'busy' trap is easy to fall into…The truth is that we are all much less busy than we think we are. And our consistent insistence that we are busy has created a host of personal and social ills."[174]

The triple-stress from passive screentime is that it

1. Artificially manufactures stress when there is nothing really stressful happening
2. Reduces our ability to handle stress
3. Steals so much time that we put things off, miss deadlines, and lose opportunities

Complaining Is a Health Hazard

We know how annoying it is to hear other people complain. But some of us who complain are not aware of how we are hurting ourselves. It undermines our health, both mentally and physically. Studies show that "complaining makes your synapses make you feel more negative and makes you keep complaining more…" it is also "weakening your immune system; you're raising your blood pressure, increasing your risk of heart disease, obesity and diabetes, and a plethora of other negative ailments."[175]

According to more studies, in addition to sabotaging long-term happiness, the habit of complaining reduces our lifespan by ten or fifteen years.[176]

While we're still alive, complaining keeps us from fixing the very problems we complain about. It is also an ugly habit, pushes other people away, and damages relationships.

Complaining happens most where there is the least emotional development. Emotional intelligence includes the tools for successful management and healthy processing of irritations, frustrations, and disappointments. This kind of intelligence comes only from nonscreen, active-minded engagement in life—when, for example, the average 21-year-old's 30,000 hours of screentime are replaced with 30,000 hours of real-life interaction and activity.

Emotional immaturity causes poor interpersonal relations, such as venting at other people's expense. Subjecting other people to rants demonstrates a self-involved disrespect for other people, and disregard for the relationship. "Emotionally intelligent people place high value on their relationships, which means they treat everyone with respect, regardless of the kind of mood they're in."[177] Emotional intelligence happens by spending time with people, instead of with screens.

By not complaining, we can actually express our problems more effectively. A complaint is a thoughtless reflex. It makes us feel worse. An expression of problems is a thoughtful conversation. It makes us feel better. It also promotes bonding.

Active minds and emotional intelligence by definition equate to a healthier personality—one that knows how to share the good and bad in life without complaining. Screentime interferes with this kind of healthy brain development. Complaining is simply a less-developed way to process the world.

How to Be Miserable

Here are ways to make sure we will be miserable. The list below is paraphrased from the article "How To Make Yourself Miserable: Discovering the Secrets to Unhappiness."[178]

- Focus on immediate results and short-term goals.
- Look for quick solutions and expect instant results without putting in any effort.
- Multitask most of the time, and never fully engage in the moment.
- Shift responsibility to others.
- Criticize others while justifying yourself.
- Often compare yourself to others.
- Feel envy.
- Practice victimology—taking the role of victim in the blame game.
- Contemplate the things someone else has done to hurt you.
- Exhibit a bad attitude about most things, events, and people.
- Don't exhibit gratitude.
- Label challenging situations as terrible or horrible.
- Spend time complaining about the way things are.
- Believe you are entitled to anything.
- Resent not getting what you think you're entitled to.

We have ample scientific evidence demonstrating that this list fits the profile of a screen-centric lifestyle, affecting almost everyone today.[179]

Physical Cost

In addition to undermining our mental and emotional development, TV undermines physical health. "There's no shortage of research showing links between watching too much television and early death.

- "Every hour of television you watch reduces your life expectancy by close to 22 minutes.

- “The researchers found people who watched more than three hours of television a day had double the risk of premature death when compared to those who watched less than one hour per day.
- “But when they looked at other sedentary behaviors—driving a car and using a computer—they didn’t find the same links with early death.”[180]

What’s special about the findings is that other sedentary behaviors don’t have the same death-correlation as watching TV. *TV specifically damages health, over and above the negative impact of sedentary behavior in general.*

Clinical psychologist and behavioral therapist Dr. Gadi Lissak assembled a report from dozens of studies conducted at research centers around the world. Among other results, he describes the physical consequences of screentime on children. The studies found that screentime is associated with risk factors for cardiovascular diseases such as high blood pressure, obesity, low HDL cholesterol, poor stress regulation, and insulin resistance. Other physical health consequences include impaired vision and reduced bone density.[181]

It gets worse. Every year TV-viewing is linked to 1,140,480 new cases of diabetes, 246,240 deaths from heart disease, and 673,920 additional deaths. “Evidence from a spate of recent studies suggests that the more TV you watch, the more likely you are to develop a host of health problems and to die at an earlier age.”[182]

- “In a new analysis published … in the Journal of the American Medical Association, researchers combined data from eight such studies and found that for every additional two hours people spend glued to the tube on a typical day, their risk of developing type 2 diabetes increases by 20% and their risk of heart disease increases by 15%.
- “The increased risk of disease tied to TV watching ‘is similar to what you see with high cholesterol or blood pressure or smoking’
- “The findings are remarkably consistent across different studies and different populations
- The studies “included more than 175,000 people around the world and generally lasted between 6 and 10 years....most controlled for a

> long list of health factors (such as body mass index, cholesterol levels, and family history of disease) in an effort to pinpoint the effect of TV watching."[183]

Again, these studies *rule out sedentariness alone*, and other factors, and *specifically pinpoint TV as the direct cause of death and disease*.

Other studies have found similar results. The *International Journal of Eating Disorders* reports that "each hour of total screentime per day was prospectively associated with 1.11 higher odds of binge-eating disorder."[184] And the journal *Pediatric Obesity* found that "examining specific screen time behaviours, each additional hour of texting and video games was significantly prospectively associated with higher body mass index (BMI) percentile."[185] The abundance of scientific confirmation leaves no question about the dangers of screentime to health. The only question is how we respond.

CHAPTER 2: CONSUMER BEHAVIOR

Our mind is shaped from earliest childhood, and our perception is formed according to our environment. When our mind is immersed in screens from a young age, perception of the world around us is molded by the medium and its messages. Like the fish-in-water analogy, we are the fish, and the water is the practically seamless array of screens surrounding us. We never stop to question where the water came from, or whether there is a world outside of it. "And the deeper our immersion becomes, the less likely it seems we'll poke our heads above the surface and see there must have been life before someone invented TV."[186] It is natural for us to perceive no other reality beyond the screen-surroundings in which we live and breathe from infancy.

Our screen experience creates our perception of reality, determines what we believe, and defines who we are. Naturally we don't want to believe we have been "brainwashed"—perhaps a more accurate way to put it is that "we are formed and conditioned by an artificial reality that replaced natural development."

We are unable to imagine a possible environment outside of this "water," in which we grew from infancy to adulthood. What about life before our submersion under today's ocean of screens? We are convinced that our perception of reality, whatever it may be, is the truth. Even though science shows us that screentime weakens our judgment, we have nothing else to compare against, and we are surrounded by others who reaffirm this normality.

Our lifelong submersion under an ocean of screens creates our reality. This unnatural alteration of perception does not necessarily mean a specific episode of mind control. Screen-centered media does not have to control our minds in any specific way to influence our behavior. Instead, full-surround screen reality reshapes our overall perception, judgment, and decision making, over decades of immersion. In other words, growing up with screen-saturated conditioning, our brain simply functions less effectively, and more predictably, than it would have otherwise.

Consumer Culture

"Consumer Culture" is a culture focused on spending money for material goods and services as its primary value. The United States has been considered the world's leading consumer culture since the 1950s. Central to this system is marketing and advertising, as Wikipedia phrases it, "enticing and shaping and even creating consumerism and needs where there has been none before."[187] The most effective medium for such marketing and advertising is the screen.

The 1950s is the pivotal decade when screens rose to prominence in the homes and lifestyles of most Americans. As PBS notes in *The Rise of American Consumerism*, "Families of all income brackets were buying televisions at a rate of five million a year ... television provided a potent medium for advertisers to reach inside American homes, creating desires for other products."[188] (TV started in 9 percent of homes in 1950, and increased to 87 percent of homes by 1960.[189])

Consumer Culture as we know it today was invented by the media. Through the 1950s and 1960s it gradually formed how we think, feel, and view ourselves. It reshaped our perception of what is necessary, versus what is optional. "TV has also helped create an increasingly consumerist and considerably helpless people who are thoroughly convinced they must continue buying more and more unnecessary things."[190]

A big part of the new consumerism hatched by 1950s TV was the teen demographic. "Television also helped to create a brand new demographic: the teenager. While teenagers were already at work forming their own subculture, advertising agencies realized that teens were a potentially lucrative group to target since they had leisure time and spending power, unlike previous generations of adolescents. So television commercials were geared toward the new demographic. Teens responded by spending their money on Coca-Cola, M&Ms, and all the other products commercials sold to them—and by influencing their parents' spending habits."[191]

Those teens of the 1950s are now grandfathers and grandmothers. They helped create the new consumer mindset that grew stronger and spread faster over subsequent generations. "The spending habits we consider normal were born in the post-war 1950s."[192] This dramatic change in values and behaviors, driven by the nascent screen culture of the 1950s, defines the American Consumer Culture that we take for granted today.

Former luxuries became perceived as necessities by most everyone over the first twenty years of TV's influence. It's embedded in our consciousness to the point that it has redefined what we view as a necessity. We have internalized Consumer Culture such that impulse buying for immediate gratification has changed from "extravagant" and "irresponsible to "expected" and "normal."

We have become truly consumers at heart. We spend more and, not surprisingly, we get deeper in debt. Today, US households owe $trillions in debt. Household debt relative to GDP has increased dramatically since those first TV ads reached that first-ever teen demographic of the 1950s.[193]

As described by author Allan Stromfeldt Christensen in his Lemminged article: "Since the mass production of consumer goods requires mass consumption to keep the machines running, television—and thus mass advertising and mass media—has played a leading role in promoting and upholding these values. CEO Lowry Mays of Clear Channel, a media conglomerate with 900 radio stations in the US, plainly stated that 'We're not in the business of providing news and information…We're simply in the business of selling our customers products.' Since television exists in a similar vein of selling what are mostly optional products, it's fair to say then that what television props up is essentially a false economy, arriving under false pretenses."[194]

Since Americans are now surrounded by screens from infancy, we become lifelong products of the media's demand-creation influence over us. We are consumers created by the media, to support the media, by increasing the revenue of the media's clients.

Growing up amidst screen media force-feeds us consumption-driven messages that replace real-life experience. Before we can possibly be conscious of it, or defend ourselves against it, we are formed into passive consumer personalities from infant to adult. "The greatest accomplishment of television then has been to pass itself off as a conveyor of entertainment, news, and occasionally ideas, since its real purpose has been to maintain an audience receptive to commercial advertising … Or rather, *TV doesn't so much advertise products as much as it promotes consumption as a way of life*."[195]

Once the whole population's perception of reality is reshaped into a consumer mindset, trends in people's decisions and behaviors will fall into place in a predictable way, statistically speaking. Corporations spend $billions on advertisements and on programming, knowing *with certainty* that a predictable percentage will respond as desired.

"By bombarding audiences with an array of subtle and not-so-subtle messages aimed at breeding dissatisfaction in themselves, their cultures and their values, the ultimate goal of television has been to convince people that human needs such as creativity, compassion, understanding, freedom, leisure and security can be replaced and satisfied through material consumption. TV has also helped convince viewers that, when their consumption doesn't seem to be delivering the promised benefits, people just need to consume more."[196]

Our identity is created and formed by the media, through the screens that surround us. We involuntarily and unconsciously become the intended creation: the easily manipulated, impulse-buying consumer who compulsively falls deeper and deeper in debt under the influence of uncontrollable urges to spend money on unnecessary goods that very soon end up in the trash. Our deepest thoughts and feelings are completely conditioned.

Green by Association

With today's ubiquitous smartphones, satellites, computers, other devices, video games, bigger TVs, more TVs, and more automobiles per household than ever—plus the rise in many other material goods—so energy consumption and carbon footprint per person have risen, far beyond the levels of past generations.[197]

Today's highest levels of consumption in history didn't just happen. They formed over recent generations, as screen-messaging conditioned us to perceive it as normal. As science has proven, screentime makes our minds suggestible, and screen-messaging encourages wasteful indulgence while it convinces us that we are more environmentally conscious than pre-screen generations. We are proud of how "green" we are today. Let's step back for a moment, and take a look at how we became oblivious to this striking contradiction between perception and reality.

One of the areas where government regulations have benefited society is in stopping the massive dumping of toxic waste, flammable waste, radioactive waste, chemicals like DDT, and other deadly pollution flowing untreated into our aquifers, rivers, lakes, and oceans. North America was a free, unregulated dumping ground throughout the late 1800s and the early 1900s.

At that same time, private citizens were less wasteful. There was no plastic. We reused jars, tin cans, fabric, wood. There was no packaging for anything as we know it today. Many behaviors that we call environmentally conscientious today, were standard default behaviors of private citizens before the 1950s.

By the mid to late twentieth century, a reversal began to happen. Environmental regulations were ramping up for businesses, just as private citizens began wasting and polluting more. Companies began promoting their sustainability efforts, reducing their carbon footprint, and other green-related policies. Corporate ads on TV touted green initiatives and environmentally friendly practices. By association, we felt good about our enlightened society *as seen on TV*. But our personal behaviors drifted further and further away from those environmentally enlightened behaviors. Today, we have in fact become far more wasteful and self-indulgent than ever before.[198]

Marketing Works

Marketing experts estimate that "Americans are exposed to around 4,000 to 10,000 advertisements each day."[199] Under the influence of those millions of ad messages from infancy to adulthood, we start believing that material goods will fulfill us, and that a walk in the park, for example, will be boring and unsatisfying by comparison. We begin to believe that what is false (that a product will fulfill us) is true; and that what's true (that experiencing nature will fulfill us) is false.

As filmmaker and blogger Allan Christensen writes, "Even if the message on TV is pro-environmental, TV viewing is intrinsically anti-environmental because it provides a substitute for experiencing nature first hand and because it encourages passivity.... In addition, because electronic media are able to make consumer products seem 'more alive than people,' it is inherently biased toward materialistic consumerism...."[200]

Screen-messaging told us that we can be environmentally responsible even while consuming more and increasing our personal carbon footprints and vastly increasing our use of resources. Screen messaging molded our mindsets to make us wasteful and proudly "green" at the same time. As passive screen viewing compromised our critical thinking, it became easy for us to embrace this remarkable duality. Some examples of our behavioral disconnect include:

- Owning more cars and driving more than ever before
- Increased consumer appetite for large-carbon-footprint luxuries such as smartphones
- Switched from glass to plastic containers, including an endless supply of plastic bags in stores (plastic is an oil product and doesn't decompose easily)

Today's waste-based consumer culture presents a much larger carbon footprint than at any time in the past.[201]

Cars and Driving

By owning more cars than ever, we have counteracted the gains from more-efficient engines. We lean on engine technology to justify automobile indulgence, hoping to just about break even. "The single largest source of emissions for the typical household is from driving (gasoline use)."[202] In fact, our carbon pollution today is double that of 1960.[203] Electric and hybrid vehicles should help in the future. But as of 2023 we use coal to produce the electricity needed to power these vehicles, battery manufacturing and disposal take a larger toll, and manufacturing the vehicles themselves has a significantly larger carbon footprint than manufacturing gas-powered vehicles.[204]

Today, kids drive or get driven everywhere they go, instead of walking or riding bicycles with friends. Kids also drive to school, or get driven by their parents. Observe an elementary, middle, or high school in the morning and watch the legions of SUVs dropping off kids. There used to be a legion of bicycles with no parents in sight. Even the wealthiest children in the past rode their bicycles, and would never have dreamed of today's commonplace luxuries and parental indulgence. Kids bicycle-riding and playing outside with no video devices were much healthier, and represented a truer model of "going green." Then came the screen messaging that conditioned our minds to accept the dissociation between perception and reality.

Internet Communications and Smartphones

Today we take Internet communications and smartphones for granted, which require building, launching, and maintaining satellites to support 24/7 entertainment for everyone. Satellites have become a necessity: "Without them, we wouldn't be able to browse the internet on our phones, send picture messages or make long distance phone calls." To communicate and to triangulate GPS information, satellites require earth-based stations and thousands of signal antennas.[205] More of all of the above are constantly in progress.

Our least-green devices, smartphones, use large amounts of advanced materials made from rare core ingredients. Smartphones alone have caused the carbon footprint of the entire technology and communications industry to triple. According to a recent report based on independent studies, "A significant reason behind this is mining of rare minerals which are required for manufacturing of new smartphones. They are representing about 85% to 95% of phone's total Carbon emission in its average 2 years of life." Customers are not doing their part either, as Lotfi Belkhir, author of an independent study notes, "based on our research and other sources, currently less than 1% of smartphones are being recycled."[206] While smartphones have become ubiquitous, and difficult to avoid, recycling them is a no-brainer. But still, 99 percent of the time we don't bother.

Green Mirage

Advanced materials manufacturing, deploying, maintaining, and upgrading today's massive land-air-and-space communications infrastructure, dependence on smartphones, negligence of recycling, parental indulgence and the new chauffeur culture, as a combined total, gives us a global carbon footprint beyond the imagination of past generations. The past seventy years brought ever-increasing screen-based consumerism and deceptive messages to appease our consciences in order to increase profits. Each generation since then has increased consumption and wastefulness.

Today, we are green only when convenient, not because we care about *being* conscientious, but because we care about *feeling* conscientious. We do what the screen says we can do, and we believe what the screen tells us to believe, and we live comfortably with the contradictions. We view ourselves as environmentally responsible people because it feels good to view ourselves as environmentally responsible people. We participate in the "green" worldview because it is a socially acceptable illusion that avoids critical reasoning. As shown in many studies referenced earlier, critical reasoning is one of the first abilities undermined by screentime.

CHAPTER 3: SOCIAL DEVELOPMENT

"No social group…can survive without constant informal contact among its members," says Dr. Christopher Alexander, in his classic work *A Pattern Language*.[207] Alexander's book highlights hundreds of detailed ways in which human interaction must be built into cities, parks, neighborhoods, houses, rooms, sidewalks, yards, gardens, public spaces, office buildings, commercial spaces, and many more. The very foundation of human survival rests in frequent face-to-face interaction, and deep in-person connections reinforced by daily renewal, in person. Screen isolation diminishes or eliminates all of the above, and erodes social foundations that underpin human health.

Society is made up of the collective abilities, practices, and beliefs of its members. A society of millions is not significantly harmed if a few hundred people drop out. But if the whole population of millions are systematically isolated from one another, the health of the whole society systematically declines mentally, emotionally, and physically.

Social Skills

As Child Psychiatrist Victoria Dunckley points out in *Psychology Today*, "The more a child hides behind a screen, the more socially awkward he or she becomes, creating a self-perpetuating cycle. In contrast, a shy child who continually works at overcoming social anxiety is likely to overcome it. … Nowadays, socially anxious or awkward children and teens *aren't forced to practice face-to-face and eye-to-eye* interaction because some of their social needs are met online." As a result, "the ability to tolerate the physical presence of others never builds, and 'walls' are erected instead."[208]

There are broader implications to the screen-related loss of social skills. Here are some typical health consequences for teens who use screens:[209]

- "social incompetence"
- "often act much younger than their years"

- “tend to make poor eye contact, seem distracted or ‘not present,’ or squirm with discomfort”
- “seem apathetic and demonstrate passive body language”
- “unable to engage in meaningful, reciprocal conversation.”
- “not be able to follow longer or more nuanced questions because of a shortened attention span”
- “They may also hold grudges or attribute hostile motives to others where there are none.”
- “have a low frustration tolerance that results in meltdowns and a tendency to blame everyone but themselves.”
- “less able to tolerate disappointment and boredom, more entitled, and less willing to work—whether it be for school, at a job, or to improve a relationship.”

The low tolerance for frustration and disappointment is partially due to the fact that the “negative impact of screentime on the brain’s frontal lobe, lack of eye contact and face to face interaction, this dynamic occurs because screen activities tend to create a false experience of ease and success: electronic media offers immediate gratification, endless (and effortless) stimulation and entertainment, the ability to control one’s environment or one’s image, and the opportunity to be a hero — features that don’t reflect how things work in the real world. *Real life is much more difficult.*”[210]

As research progresses, our understanding of screentime-related damage only grows more alarming. As the founder of San Diego’s Center for Mental Health and Wellness, clinical psychologist Dr. Azmaira H. Maker writes: “There has been a significant amount of research conducted on this topic in recent years that shows the following evidence:

“The Social and Emotional Effects:

- Increase in stress
- Increase in time to complete tasks
- Increase in off-task time
- Increase in anxiety with no access to electronics
- Increase in frustration and decrease in commitment to deeper, more challenging tasks and problem solving
- Increase in impulsivity

- Decrease in emotional regulation
- Decrease in ability to recognize facial emotions and non-verbal cues

"The Neurobiological Effects on the Developing Brain:

- Repeated release of dopamine, increasing pleasure and addiction
- Chronic need for stimulation and instant gratification
- Decrease in focus and attention span
- Increase in arousal
- Blue light - Shut down of the pineal gland that releases melatonin (a natural hormone to induce sleep)
- Sleep deprivation: poor sleep and less sleep
- Sensory overload

"Screen Addiction can also lead to:

- Grey matter shrinkage (where processing occurs)
- Frontal lobe shrinkage (where executive functioning occurs, such as planning and organizing)
- Striatum shrinkage (where reward pathways and impulse control of socially unacceptable behaviors occur)
- Insula damage (where our capacity to develop empathy and compassion occurs)
- Loss of integrity of white matter (these are the connective pathways for communication within the brain)
- Impaired cognitive functioning
- Reduced number of dopamine receptors, which is linked to depression"[211]

In summary Dr. Maker notes that "The research strongly suggests that human to human, hands on interaction is the most beneficial for a child's socio-emotional development, as screen time could impair empathy, communication, cognitive functioning, emotional regulation, sleep, attention, and brain development."[212] Add to this the fact that screentime leads to increased likelihood of aggressive and criminal behaviors.[213] It is hard to overestimate the magnitude of the sociological impact—the development, structure, and functioning of human society—with a whole population so affected.

Note that these are consequences over and above naturally occurring issues in these areas that children might have already. Whatever the socio-emotional condition of children, no matter how poor the child's mental health is already, screentime will make it worse. If the child didn't have these problems before, they will get them, and if they already show these symptoms, they will get worse with every hour of screentime.

Thus while we as parents think our kids are just having fun with the latest toys, their basic brain functions are being suppressed. With every hour of screentime, they become less competent, less functional human beings, whose chances in adulthood are being ruined by our failure to support their well-being.

Conversely, teens who give up screens will experience "dramatic leaps in maturity in terms of conversational skills, eye contact capacity, and empathy or insight."[214]

Even without screentime's known damage to intelligence, creativity, and physical health, these disastrous social and emotional consequences for our children and our society should be ample warning to ban screens in our children's lives while they are not yet old enough to know better.

Each Mind Is a Building Block of Society

Each mind of each person in a society is a building block of our social fabric. A population of active-minded people achieves a healthy society and a vibrant culture. Such a population doesn't just happen by itself. It requires everyone's effort and development. Active people make it happen.

Unfortunately, there is an inverse relationship between passive screentime versus active development. As screentime goes up, development goes down. Culture deteriorates because the building blocks of culture deteriorate. We are the crumbling building blocks.

Extend screentime's impaired development of individuals across the mass population. Development of every area of everyone's mind is undermined. Here are some of the building blocks of our minds and of our culture that we have damaged by simply spending time with screens[215]:

- Verbal skills
- Math skills
- Intelligence (IQ)

- Emotional intelligence (EQ)
- Maturity
- Imagination
- Attention span
- Concentration
- Critical processing
- Analytical thinking
- Bonding
- Social skills
- Motor skills
- Mechanical skills
- Physical condition
- Overall fulfillment
- Anger and conflict management
- Stability
- Responsibility
- Reliability
- Initiative
- School and work performance
- Ability to hold a job
- Future planning
- Delayed gratification

The greater number of screen hours, the greater the deficiency in all of these areas.[216]

After generations of increasing screen habits, social deterioration followed with inevitable force. It's no surprise that drug abuse and crime increased suddenly and dramatically in the mid-twentieth century. Today's reality of approximately 340,000,000 people suffering these same declines en masse is a recipe for the spread of destructive tendencies like drug abuse and crime.

As noted earlier, TV started in 9 percent of homes in 1950, and increased to 87 percent of homes by 1960.[217] Experts hypothesize all kinds of explanations for drug abuse and crime, but they only address symptoms. The spread of TV alone makes criminal behavior more likely, according to a study reported in "Children who watch 'excessive' amounts of TV are more likely to have criminal convictions, exhibit aggression and experience negative emotions: study."[218]

Aggression comes from a passive mind, not an active mind. A population watching TV doesn't mean everyone becomes a criminal. But it means more people will become criminals. "With every hour in front of the television, kids were more likely to show aggressive behavior or receive a criminal conviction by early adulthood."[219] The rest will just become less healthy people.

Caring for Others

Screentime reduces caring for others with mathematical certainty. Lack of concern for others is a direct result of less contact with others. Less contact causes less familiarity with people, and therefore, our brains do not produce the feelings of intimacy and camaraderie. This fact is reported in a recent *Scientific American* study which concluded that quantity of in-person interaction correlates to the amount we care about others.[220] Caring is not a learned moral principle. It is a biological as well as psychological process, which happens slowly over years spending thousands of hours with people. The moral code is an after-the-fact idea that follows from how we actually feel.

As San Diego State University Professor of Psychology Jean Twenge points out, "the number of teens who get together with their friends nearly every day dropped by more than 40 percent from 2000 to 2015; the decline has been especially steep recently. … The roller rink, the basketball court, the town pool, the local necking spot—they've all been replaced by virtual spaces accessed through apps and the web."[221]

The declines in social time that seemed so alarming from the 1960s to 2000 seem negligible by comparison with iGen's virtual solitary confinement.

So it is an inevitable change in human nature: as contact with others decreases, so caring for others decreases. As isolation increases, so self-centered thoughts, feelings, and behaviors increase.[222] Time with other people increases our emotional maturity: More time with others makes us focus on others. More time alone makes us focus on ourselves.

Focus on others makes us better listeners in conversations. We want to ask questions, we show interest. It is not a formula, it is a natural instinct and a true feeling. Time increases interest. We want to get to know each other better, to share in others' experiences.

As *Emotional Intelligence 2.0* author and clinical psychologist Travis Bradberry notes, "To be deliberately empathic, you have to let your ability to walk in their shoes change what you do, whether that's changing your behavior to accommodate their feelings or providing tangible help in a tough situation."[223]

Genuine caring on the inside (as opposed to an "act") goes hand in hand with sharing experiences with one another. It means understanding and caring for people with whom we disagree on life choices, religion, politics, social issues, musical taste, and most any issue.

Again, Dr. Bradberry puts it succinctly: "To eliminate preconceived notions and judgment, you need to see the world through other people's eyes. This doesn't require you to believe what they believe or condone their behavior; it simply means you quit passing judgment long enough to truly understand what makes them tick."[224]

This true sense of caring and understanding does not develop in people who lead a screen-centric life. It *cannot* grow where screentime has replaced people time. A screen-dominated personality lacks empathy, which means we cannot understand "caring" as it was once felt.[225]

Caring for others is the *nature of being* for people who grew up spending a lot of time with others. Caring has become another screen-culture casualty.[226]

From Meaningful Philosophy to Wealth and Glamour

The meaning of life and fulfillment represent another casualty of screentime. For example, screen-messaging has changed us from having deeper life values—which would provide authentic fulfillment—to valuing wealth for the sake of being rich, which leaves us empty. Screen-messaging creates and perpetuates material wealth as our number-one cultural value.

Recent research revealed that the "percentage who say it is 'essential' or 'very important' to be 'very well off financially' grew from

41.9% in 1967 to 74.5% in 2005; 'developing a meaningful philosophy of life' dropped in importance from 85.8% in 1967 to 45% in 2005." Additionally, 81% of 18- to 25-year-olds surveyed in a Pew Research Center poll said getting rich is their generation's most important or second-most-important life goal." The *young people themselves* cited the influence of TV shows promoting wealth and glamour as the factors in their values and outlook.[227] Results of another study confirm that smartphone use predicts adolescent materialism.[228]

The elevation of materialism caused by screentime among GenZ is more than just a conceptual shift in values. Psychologist Tim Kasser's research shows that the more materialistic people are—the more they think about acquiring wealth, or stuff, or fame—the more unhappy, depressed, and anxious they will be. Teens become anxious, depressed, lonely, and feel empty inside. Empirical data over many years reveals that valuing wealth and possessions face a greater risk of unhappiness, anxiety, depression, low self-esteem, and problems with intimacy. These outcomes occur regardless of age, income, or culture. Material motivation makes us feel less free and more insecure.[229]

The fact that passive screentime systematically lowers brain function has been clearly proven in every imaginable area of brain activity.[230] This added dimension of anxiety and depression only deepens the emptiness.

Irritable TV Viewers

In families, members grow more distant from one another when they spend time on their separate devices—even when they watch TV "together." Home becomes a colder place. Family members are not in touch with other people in the room. They are not interacting.

Viewers are easily irritated when someone in the room starts talking. This is because of the Alpha/Beta Wave switch that we learned about in an earlier chapter. The brain shifts to sleeplike Alpha Waves while watching TV, and must be jarred back into Beta Waves to engage with the person talking.[231] The shift irritates us when we are in the sleep-like Alpha state, like an unwelcome alarm clock going off. Instead of greeting a loved one entering the room, we snap at them for disrupting our hypnotic TV-viewer state of passivity. Hence, watching TV together isolates individuals, divides families, and engenders negative feelings towards one another. It is not a way for families to bond.

A Laboratory of Screens

Our society has become like a laboratory of screens. We are the subjects—our mental and physical health being observed, studied, and documented as it steadily deteriorates (such as all the studies cited in this book).

Short term screen-fed rewards, such as dopamine hits, reinforce our immediate-gratification impulse. The addictive, brain-disabling cycle prevents us from noticing the side-effects of mental and physical damage. This behavioral cycle is similar to that of experimental laboratory animals such as in the behaviors of drug-addicted lab mice. We are left in this helpless state, wanting more and getting less.

The unintentional experiment on the effects of screentime has reduced us to complete unawareness of our gradual deterioration. We believe we are in control of our own values and decisions, while repeated reinforcements delivered via screentime guide our perception and our behavior. But these effects are not entirely unnoticed, and not entirely unintentional. As reported in the earlier War on Kids section, these effects are studied and leveraged by the makers of screen-based entertainment, in the form of persuasive technology.

As child psychologist Richard Freed reports, "Revealing the hard science behind persuasive technology, Hopson [Video game developer John Hopson, Ph.D.] says, 'This is not to say that players are the same as rats, but that there are general rules of learning which apply equally to both.'"[232] By comparing human behavior to the behavior of lab rats, video game developers successfully achieve maximum customer addiction.

To reiterate from the War on Kids section, with persuasive design in screens and apps, "around the world hundreds of thousands of people are going to quietly change their behavior in ways that, unbeknownst to them, feel second-nature but are really by design.' Programmers call this 'brain hacking,' as it compels users to spend more time on sites even though they mistakenly believe it's strictly due to their own conscious choices."[233]

"Dr. Freed and 200 other psychologists petitioned the American Psychological Association in August to formally condemn the work psychologists are doing with persuasive design for tech platforms that are designed for children."[234]

As we saw extensively reported in the earlier War on Kids section, "If you haven't heard of persuasive technology, that's no accident—tech corporations would prefer it to remain in the shadows, as most of us don't want to be controlled and have a special aversion to kids being manipulated for profit. Persuasive technology (also called persuasive design) works by deliberately creating digital environments that users feel fulfill their basic human drives—to be social or obtain goals—better than real-world alternatives. Kids spend countless hours in social media and video game environments in pursuit of likes, 'friends,' game points, and levels—because it's stimulating, they believe that this makes them happy and successful, and they find it easier than doing the difficult but developmentally important activities of childhood."

Studies cited in this book have provided plentiful scientifically verified evidence that the population has been damaged, weakened, molded, corrupted, and in many cases killed, by screentime—and that we are largely unaware of this correlation or this damage. Screen-addiction and screen-induced passivity have spread across hundreds of millions of people over several generations—making us less-capable, less-mature, less-stable, less-healthy—a predictable, pliant, less differentiated people. We didn't notice it, and we don't believe it.

We enact screen-guided behavior as we lose the capacity for critical self-knowledge, because those functions are being systematically disabled. The result is an incremental deterioration of subjects that are increasingly spellbound within a maze of screens.

Being Informed

Words of warning do not reach the public because we rely on the culprit itself for our information. If what we learn comes from screens, we will not learn about screen-related damage. Screen-based information creates the illusion of being informative, while it feeds "guided reality" into minds made suggestible by the medium itself.[235]

As noted scholar and *The Death of Expertise* author Tom Nichols points out: "Just as clicking through endless Internet pages makes people think they're learning new things, watching headlines is producing laypeople who believe—erroneously—that they understand the news. Worse, their daily interaction with so much media makes them resistant to learning anything more that takes too long or isn't entertaining enough."[236]

The more we use screen-sources to be informed, the less informed we are. The screen itself makes us less aware of the deficiency.

Taking Action

Changing our behavior is the only way to change the consequences. For example, damage from screen-based Alpha Wave passivity can be reversed, but only by eliminating or significantly reducing screentime. This kind of lifestyle change has been rewarding for many people, and worth the struggle. But it can be very difficult at first.

For example, in a Kubey and Csikszentmihalyi study "Television Addiction is no mere metaphor": At first, where families try eliminating TV, "family members had difficulties in dealing with the newly available time, anxiety and aggressions were expressed.... People living alone tended to be bored and irritated."[237] These are typical symptoms of addiction withdrawal: anxiety, aggression, irritation. But after the first few weeks, symptoms are reduced and people begin leading healthier lives, mentally and physically. The same process applies to smartphones and other screen addiction.

Like the rehabilitation from any addiction, the very idea can generate waves of anxiety. Think of the idea of getting rid of your TV and your phone. What would you do? Such a loss seems overwhelming to most. Each of us must decide for ourselves: Is it worth the trouble?

Even if it's not worth it for yourself, you can give your children a huge advantage in life by keeping them screenfree for their first six or seven years. They will be lightyears ahead developmentally. And they won't even notice the absence of screens all round them, because they never saw a screen. To them, it's no hardship, it's just a natural childhood, as natural as it was for the past 200,000 years until recently. They will thank you.

CHAPTER 4: ANGER AND HOSTILITY

One of the most obvious lifestyle differences between the past and the present is the reduction in physical activity. Past lifestyles required much more physical engagement for everyday activities in the home, at work, and at play. Today's softer lifestyles combine TV viewing, other screentime, convenience from appliances and other technologies, and more sedentary types of jobs than ever before. It is inevitable that men today would be physically weaker than men in the past.[238] Our bodies are made for activity, and today our bodies are not getting it.

Exercise Regularly

There are endless, seemingly infinite, numbers of studies showing us how important it is to exercise regularly and rigorously. Recent findings add that aerobic exercise actually renews the brain. According to Karen Postal, president of the American Academy of Clinical Neuropsychology, studies show that "new neurons are produced in the brain throughout the lifespan, and, so far, only one activity is known to trigger the birth of those new neurons: vigorous aerobic exercise."[239]

For so many reasons, thirty minutes a day of focused painful exercise must be part of a healthy lifestyle. It provides deeper peace of mind, a more balanced flow of serotonin and other healthy brain chemicals, a stronger more focused mind, increased clarity and alertness, less chance of heart attack or stroke, increased sexual ability, better circulation, reduced toxins, reduced likelihood of diabetes, reduced likelihood of cancer, stronger immune system, better and deeper sleep every night (again boosting immunity), easier time waking up every morning, and less physical discomfort doing everyday activities—plus a happier, stronger, sharper mind.[240]

As *Emotional Intelligence 2.0* author and clinical psychologist Travis Bradberry notes, "None moreso than vigorous exercise … releases chemicals in your brain like serotonin and endorphins that recharge it and help to keep you happy and alert. They also engage and strengthen areas in your brain that are responsible for good decision-making, planning, organization, and rational thinking."[241]

In addition to specific routines, we can supplement formal exercise with choices such as taking the stairs instead of the escalator or elevator, if at all possible. Taking the stairs improves brain function and neuronal health ("brain age decreases…by 0.58 years for every daily flight of stairs climbed").[242] While taking the stairs is not possible for many people, for those who are fortunate enough to have that choice, it is foolish not to take it.

The easy path of inactivity yields higher anxiety, deeper depression, fewer brain neurons, less energy, a weaker less-focused mind, dementia, unbalanced brain-chemical flow, reduced sexual life, greater risk of heart attack and stroke, increased likelihood of cancer and diabetes, reduced circulation, weaker immune system, troubled sleep, a harder time getting out of bed every morning, disorganization, irrational thinking, and more physical discomfort doing everyday activities.

Brain exercises also enhance brain health. One particular point proven in recent years is that mental exercise prevents memory loss, Dementia, Alzheimer's, senility, and mental deterioration in general. Conversely, TV viewing is an indicator of these diseases.[243] The main obstacle keeping most of us from an active healthy lifestyle is screen addiction. It creeps into our lives and makes us mentally and physically sick.

Emotional Instability and Anti-Social Behavior

In addition to diseases and disabilities, our screen-based lifestyle makes our emotions less stable, causing unhealthy amounts of impatience and anger. A passive mind yields an irritable personality. It reduces our ability to effectively process irritations and aggravations, both large and small.

Processing negative experiences takes effort. Screentime weakens our minds so we cannot take that effort. We cannot process experiences effectively, so we become short-tempered, easily upset, irritable, snapping at people, and self-centered. This stunted emotional development is precisely what a University of Otago study found in children growing up watching too much TV. The research, tracking 1,000 children over a ten-year period, found a strong correlation between TV and anti-social behavior.[244]

Short-temperedness typically results from bitterness and stress. This unhealthy combination signals our failure to develop Emotional Intelligence (EQ). Those who do not process anger very well, suffer greater stress. As clinical psychologist Travis Bradberry notes, "Researchers at Emory University have shown that holding onto stress contributes to high blood pressure and heart disease. Holding onto a grudge means you're holding onto stress, and emotionally intelligent people know to avoid this at all costs."[245]

Yet another scientific study in 2018, which included a random sample of 40,337 2- to 17-year-old children and adolescents in the US, showed that increased screen use led to loss of self-control and emotional stability.[246]

Holding on to unresolved bitterness, stress, and anger, we are pushed around by events instead of processing events in a way that gives us command of the moment. Failure of emotional intelligence makes us appear immature, irritable, or self-centered, but the root cause is a passive mind. The symptoms include the tendency to snap at others, slam doors, act upset without serious provocation, and generally exhibit self-involved behaviors. Emotional reactions such as these demonstrate an exaggerated need for self-gratification and lack of being in touch with the outside world. These behaviors push others away, and damage relationships. As Dr. Bradberry explains, "Expressing anger too much or at the wrong times desensitizes people to what you are feeling, making it hard for others to take you seriously."[247]

The tendency to lose self-control increases as each new generation grows up immersed in more and more screens. Today there is a huge gap in Emotional Intelligence between the generations. Dr. Bradberry's book reveals, for example, that Generation Y (bn. late 1980s–90s) are much more prone to fly off the handle than previous generations when things don't go their way.[248] As these behaviors become the norm, they redefine the culture. Screen habits produce and reinforce this behavior. Self-gratifying anger increases when the culture condones it.

As Child Psychiatrist Victoria Dunckley points out in *Psychology Today*, there are serious emotional-health consequences for teens who use screens. Here are just a few areas where screentime damages emotional intelligence in teens:[249]

- "often act much younger than their years"
- "tend to make poor eye contact, seem distracted or 'not present,' or squirm with discomfort"
- "seem apathetic and demonstrate passive body language"
- "unable to engage in meaningful, reciprocal conversation."
- "not be able to follow longer or more nuanced questions because of a shortened attention span"
- "They may also hold grudges or attribute hostile motives to others where there are none."
- "have a low frustration tolerance that results in meltdowns and a tendency to blame everyone but themselves."
- "less able to tolerate disappointment and boredom, more entitled, and less willing to work—whether it be for school, at a job, or to improve a relationship."

Dr. Dunckley notes one of the reasons for the above, that "screen activities tend to create a false experience of ease and success: electronic media offers immediate gratification, endless (and effortless) stimulation and entertainment, the ability to control one's environment or one's image, and the opportunity to be a hero—features that don't reflect how things work in the real world. *Real life is much more difficult.*"[250]

When our minds are passive, confronted with real life where we don't have screen-style control over our circumstances, we collapse psychologically. We become weak, emotionally, psychologically, mentally. Lack of control opens the door to reflexive outbursts. We express any negativity we may be feeling at the moment, without any process in-between the feeling and the reflex. Thus we are pushed around by external forces instead of taking command of the moment. Mindless yelling whenever we're angry shows the failure of development.

We have the power to change this seemingly hopeless condition.[251] But first we have to be aware that it's a problem, and that there is a better way.

Self-Control

Dr. Bradberry offers an insightful perspective on the culture of self-control: "One of the huge fallacies our culture has embraced is that feeling something is the same as acting on that feeling, and that's just wrong, because there's this little thing called self-control. Whether it's helping out a co-worker when you're in a crunch to meet your own deadline or continuing to be pleasant with someone who is failing to return the favor, being considerate often means not acting on what you feel."[252] "If you grow up in a culture where emotional outbursts and careless self-gratification are not only discouraged but are also considered personally shameful, such an upbringing is going to affect the way you manage yourself and others."[253]

Non-screen lifestyles produce effective self-management, taking command of our moods in a responsible fashion. "Emotionally intelligent people … treat everyone with respect, regardless of the kind of mood they're in."[254] Bradberry explains a major benefit, "Those who use the right tools and strategies for harnessing their emotions put themselves in a position to prosper."[255]

Subjective versus Objective Emotional Intelligence

There is a significant "subjective-versus-objective" component to emotional intelligence. The self-gratifying reflex is symptomatic of being excessively subjective. It leads to being controlled by emotions with negative consequences. Consciously shifting to a more objective way of processing helps break this cycle of helplessness. "Your objectivity would allow you to step out from under the control of your emotions and know exactly what needed to be done to create a positive outcome."[256]

We can take command of the moment, and transform negative emotion into a different kind of energy. "While it's impossible not to feel your emotions, it's completely under your power to manage them effectively and to keep yourself in control of them. When you let your emotions overtake your ability to think clearly, it's easy to lose your resolve."[257]

By being more objective, we also gain clearer insight into self-improvement opportunities. For example, we are better equipped to benefit from criticism, by taking it thoughtfully, instead of being offended or reacting resentfully to criticism.

Emotionally intelligent command of the moment includes a balanced response to criticism. For example, in an article about influential people, one characteristic is how to take criticism. When criticized, they "don't react immediately and emotionally. They wait. They think. And then they deliver an appropriate response. [They] know how important relationships are, and they won't let an emotional overreaction harm theirs. They also know that emotions are contagious, and overreacting has a negative influence on everyone around them." Similarly, they "do not react emotionally and defensively to dissenting opinions—they welcome them. They're humble enough to know that they don't know everything and that someone else might see something they missed. And if that person is right, they embrace the idea wholeheartedly because they care more about the end result than being right."[258]

Active-minded people are open to criticism, open to opposing views from people they dislike, objective in their response to new information, not concerned about "being right," and not defensive. Greater emotional development allows more genuine openness to criticism and opposing viewpoints. It also allows stronger character-building and more advanced personal development.

Active screenfree lifestyles lead to both stronger and more genuine personalities. "Genuine people have a strong enough sense of self that they don't go around seeing offense that isn't there. If somebody criticizes one of their ideas, they don't treat this as a personal attack. There's no need for them to…feel insulted…. They're able to objectively evaluate negative and constructive feedback, accept what works, put it into practice, and leave the rest of it behind without developing hard feelings."[259]

Objectivity is the key. Whether my mood is good or bad, being aware that it's just a mood, retaining an objective view of what I am going through, gives me better command, judgment, freedom, and success. As concentration-camp survivor Victor Frankl said: "Between stimulus and response there is a space. In that space is our power to choose our response. In our response lies our growth and our freedom."[260]

"Victor Frankl was not only a concentration camp survivor during the Holocaust, but also someone who went on to help others find goodness and meaning in life. He was a man from whom we can learn something about what it means to be human and how to be our best—sometimes in spite of our inclinations."[261]

This is the hallmark of an active mind—it overrides weaker passive emotional reactions. It is another ingredient in a higher caliber lifestyle. It is the kind of higher functioning that is another casualty of screentime.

Passive Minds and Violence

We expect a mature person to manage aggravations in life effectively. As part of growing up, we learned to restrain immediate reactionary impulses. Active minds process experiences in a way that achieves a constructive outcome. This is learned through years of conflicts and negotiations with other people, which includes being able to put ourselves in the shoes of the other person. These learnings do not happen in a screen-dominated childhood. Active processing never develops, and the brain is restructured into a more passive state of being.

A passive mind is undisciplined and therefore unwilling and unable to process stress. Passivity bottles up frustration, oppressed feelings, and unmanaged anger, which produces erratic outbursts. The untrained mind does not plan for the future, but only reacts to pain-impulses, seeking immediate-gratification, which can lead to acts of violence and eruptions of pent up frustrations. These types of emotionally undeveloped reflexive acts have no reasonable or apparent provocation.

Recent studies reveal that unprovoked, out-of-proportion violence against others becomes more likely in the absence of healthy social time. It has been shown that minds made passive by watching too much TV carry more aggression and are more likely to commit crimes.[262] Correspondingly, "A recent study finds a decline in empathy among young people in the U.S."[263] When life is mostly one of isolation from people, there is no understanding or caring for others.

Scenarios of out-of-proportion violence related to screentime typically involve children and young adults. Here are two recent examples:

1. "A 10-year-old boy fatally shot his mother for refusing to buy him a new Amazon virtual reality headset" In custody, the boy demanded his tablet and laptop. He is being tried as an adult and faces sixty years in prison.[264]
2. In 2021 a 23-year-old living at home told his parents he was taking online classes and working remotely for an insurance company. A few months later he claimed he had gotten a new job at SpaceX, and would be moving to Florida to join the company. In reality, he was spending days, months, and years playing video games. He had never worked at an insurance company, there was no SpaceX job, and he had failed out of community college after just one semester. When his parents caught on to his lies, the young man shot both of his parents to death.[265]

We have seen these symptoms in dozens of studies referenced in this book. Being brought up with video games and other screen-based entertainment is proven to significantly increase risk of all of the behaviors in this story: not completing assignments, failing in school, unable to hold a job or support oneself, continuing to live with parents, mental atrophy that intensifies anger as well as intensifying screen addiction, lying to cover screen addiction, denial, aggressive and criminal behaviors. They don't all end in murder, but they almost always end badly. Society today is rife with screentime-related mental and physical damage and the consequent behaviors.

It is important to note here, that the violent tendency is from screentime of any kind, regardless of programming. Failing to recognize that the problem is with the screen, and not what is on the screen, we will not be able to confront the root cause of many of our social ills today.

Fighting to Reduce Violence

Historically, violence has been primarily associated with male behaviors, linked to the male hormone testosterone and evolutionary survival needs (hunting, defending, etc.). But not all types of violence are equal. According to studies that focused on male socialization, healthy fighting during childhood is linked to men's enhanced mental health and maturity.

Boys who grow up to be adults without this natural formative interaction will grow up unhealthy, with less emotional intelligence, less mature self-governance, ill-equipped for healthy confrontation during conflicts in later life, developing less camaraderie and less connection to others. Men evolved to need this activity in order to develop normally.

For example, an Iowa State University study found that TV viewing is linked to ADHD[266], and in an ADHD study, "Intriguing research by neuroscientist Jaak Panksepp shows that giving young, hyperactive lots of opportunity to do play fighting helps them learn to inhibit their behavior."[267] Combine this with the proven correlation between TV viewing and criminal behavior[268], and we begin to see a troubling picture of unhealthy suppression of fighting, which undermines mental stability, and increases risk of criminal behavior.

Adding to the male-bonding-through-fighting concept, a study reported in *The Oxford Handbook of the Development of Play* found that the subtleties of fighting are typically male specific. Showing videos of healthy fighting versus serious or vicious fighting, men could tell the difference, women who grew up with brothers could tell the difference. But women who hadn't grown up with brothers could not tell the difference—they mostly thought all of the videos equally involved serious fighting.[269]

With tendencies towards traditionally "feminine socialization" today, as shown in several recent reports[270], and changing behavioral expectations, we see more blanket condemnation of boys' fighting in schools and neighborhoods. Healthy fighting becomes viewed the same as unhealthy or vicious fighting, seen as indistinguishable and wrong. So we suppress one of the most vital developmental processes for men, a process that allows men to manage conflict more effectively later on.

"Adults, especially women who aren't personally familiar with rough play often try to stop rough-housing because they don't want anyone to get hurt. But research tells us that, overall, rough play turns into a real fight only about 1% of the time among elementary school boys."[271] According to the president of the National Institute for Play Dr. Stuart Brown, "the rough-and-tumble play of children actually prevents violent behavior; that play can grow human talents and character across a lifetime."[272]

Throughout history, boys by the age of 10 would have been in many fights and wrestling matches, on the playground, in the neighborhood, wherever kids go. It happens less nowadays, partly because of changing trends in socialization[273], which can have a damping effect on some traditional male behaviors such as fighting.[274]

The reduction in rough play or non-vicious fighting also follows from increased screentime. Screens have undermined social development in general, including boys fighting. Screens damage our children in these areas of socialization, resulting in adult males less capable of managing conflict.

A study reported by the *Child Trends Databank* found that "The share of students in grades 9 through 12 who had been in at least one physical fight in the past year declined from 43 percent in 1991 to 33 percent in 2001. The proportion remained steady until 2011, between 32 and 36 percent. However, between 2011 and 2015 the proportion decreased markedly, from 33 to 23 percent."[275] The years 2011 to 2015 correspond to the years that smartphones became universal. We increased from too much screentime up to 2011, to way too much screentime by 2015. With continuous screentime, natural behaviors such as boys fighting became suppressed even more than in previous years. Increased screentime for boys leads to deterioration of healthy masculine development, and can result in what today is often termed toxic masculinity—hyper-aggressive, overbearing, and rage-style violence.

As a society we are better off if we reduce or eliminate screentime for our children, and let natural healthy development take its course. Conversely, leaving current levels of screentime unchecked will lead to more aggression, criminal behaviors, and out-of-proportion violence.[276]

CHAPTER 5: EDUCATIONAL DEVELOPMENT

Recently a team of scientists conducted a systematic review of fifty-eight studies worldwide from 1958 to 2018 researching the effects of screentime on educational development. The studies included 480,479 participants ages four to eighteen. The research individually studied the effects of television, computers, video games, internet, mobile phones, and overall screentime. Reported results included school grades, academic achievement, performance on tests, academic failure data, and other school performance.

Some key findings included the following: As screentime increased, academic performance decreased. Breakdown of the data showed that television-viewing time was specifically linked to poorer mathematics and language performance as well as overall academic performance. The findings also ruled out the reverse explanation—disproving the notion that children with developmental delays might receive more screentime to manage their behavior. The studies confirmed that screentime leads to developmental delays, not the other way around.[277]

Since the 1960s, teachers have increasingly compensated for students' screen-related cognitive impairments and shorter attention spans by trying to "make learning fun." Teachers are supposed to "get students excited" about learning. We have to "sell" it to school kids, get their "buy in," and beg them to play the game. This response does not nurture the child, it nurtures their deficiencies. The educational system today nurtures the very impairments that school is supposed to address. The tragedy is that this compensation-response only exacerbates students' risk of failure in later life.

Computers and Schools Don't Mix

To borrow from an earlier section on tech leaders, Silicon Valley private Waldorf schools do not allow computer screens in the classroom. The school even discourages students from using screen devices at home. Interestingly, 75 percent of students' parents in this anti-computer, anti-screen school are high-level silicon valley tech executives. The school, and the parents, know that "computers inhibit creative thinking, movement, human interaction and attention spans."

While the public schools naively fill classrooms with computers, the center of the nation's technology says "computers and schools don't mix." Screens have no rôle in education. "Teaching should be a human experience … Technology is a distraction when we need literacy, numeracy and critical thinking." Instead of screens, Silicon Valley Waldorf schools are filled with blackboards, colorful chalk, bookshelves with encyclopædias, wooden desks filled with workbooks, and No. 2 pencils. The outcome is success: 94 percent of the students go on to college. Parents say the rigorous training required for teachers brings a "strong sense of mission that can be lacking in other schools."[278]

Compare screenfree Waldorf school outcomes against the dreadful outcomes of public schools today. The stark contrast highlights the terrible mistake we have made: We fell for the screen-fed illusions about the benefits of technology in our schools. Our public schools are still falling for it.

As noted in an earlier section, when you see socio-economic high-end schools doing one thing and getting excellent results, and the economically disadvantaged doing a different thing and getting terrible results, we should pay attention to that "thing" whatever it is: In this case, the use/nonuse of screens. The pattern keeps repeating itself: Each new study, expecting to discover benefits of technology in the classroom, discover the opposite is true: *The Rich Get Smart, The Poor Get Technology!*[279]

Decline of Science

An exhaustive study of the recent history of science reveals a sharp decline in scientific innovation over the past sixty years,[280] correlating to the trajectory of increased screentime during that same period across the population.

The study was conducted by Professor Erin Leahey of the University of Arizona, and by science analyst Michael Park and Professor Russell Funk, both of the University of Minnesota's Carlson School of Management. The study covered six decades, analyzed data on 45 million papers and 3.9 million patents from six large-scale datasets. They found a narrowing in the use of previous knowledge along with substantial declines in science breakthroughs. Science in the past sixty years is less likely to connect disparate areas of knowledge, less likely to push science and technology in new directions, which means a significant drop in innovation across fields.[281]

Leahey and her colleagues observed substantial declines, for example, in scientific papers: "the decrease (1945–2010) ranges from 76.5% (social sciences) to 88% (technology); for patent titles (Fig. 3d), the decrease (1980–2010) ranges from 32.5% (chemical) to 81% (computers and communications). For paper abstracts, the decrease (1992–2010) ranges from 23.1% (life sciences and biomedicine) to 38.9% (social sciences); for patent abstracts, the decrease (1980–2010) ranges from 21.5% (mechanical) to 73.2% (computers and communications)."[282]

Other studies highlighted earlier have shown that screentime undermines the ability to connect disparate areas of knowledge, suppresses creativity and innovative thinking, reduces concentration and attention spans, and damages overall intellectual capabilities. The decline of science across fields, corresponding exactly to the increases in screen usage, makes sense. Scientists are human like everyone else. They grew up like everyone else spending thousands of hours watching screens instead of developing. Screen-based damage to brain function impacts the scientific community as it does all other professions and skill sets across all industries and all walks of life.

Decline of Reasoning

The ability to reason is not automatic. Mental development requires strenuous exercise just as muscle development requires strenuous exercise. As educational psychologist Dr. Jane M. Healy found in television-related research, each year students had shorter attention spans, were less able to reason analytically, to express ideas verbally, and to handle complex problems. "The effects of these universally noted trends have begun to show up even in highly selective colleges, as professors find they must water down both reading and writing assignments as well as expectations for analytic reasoning." Healy notes with these declines, students were watching television instead of reading in their spare time. When they did read, students were less able to comprehend, remember, and apply what was read.[283]

These developmental areas are increasingly impaired by screentime with each passing year, due to television, and even moreso since smartphones. As we saw earlier in studies pointing to the decline of reasoning in our universities, "A 2011 University of Chicago study found that America's college graduates 'failed to make significant gains in critical thinking and complex reasoning during their four years of college.'"[284]

There is a lot of corroboration of these trends. "A November 22, 2016 study published by the Stanford Graduate School of Education found 'a dismaying inability by students to reason about information they see on the Internet ... Students for example, had a hard time distinguishing advertisements from news articles.'"[285] The lead author of the report, Stanford University Professor of Education Sam Wineburg, said that "'Many people assume that young people fluent in social media are equally perceptive about what they find there—but the opposite is true.'"[286]

Noted by Maryanne Wolf, Director of the Center for Reading and Language Research, and Professor of Child Study and Human Development at Tufts University, our screen habits undermine "our most important intellectual and affective processes: internalized knowledge, analogical reasoning, and inference; perspective-taking and empathy; critical analysis and the generation of insight."[287]

In another study, "The research, which included both elementary school-age and college-age participants, found that children who exceeded two hours per day [of TV] ... were 1.5 to 2 times more likely to be above average in attention problems ... Brain science demonstrates that the brain becomes what the brain does...ADHD is a medical condition, but it's a brain condition ... environmental stimuli can increase the risk for a medical condition like ADHD in the same way that environmental stimuli, like cigarettes, can increase the risk for cancer."[288]

TV is exactly the environmental stimulus that increases the risk of harmful brain impairments that compromise educational outcomes.

In other research, "the study found that for every extra hour of TV a week the two-year-olds watched there was a 6 percent decrease in math achievement … a 7 percent decrease in classroom engagement. Each extra hour also corresponded with 9 percent less exercise, consumption of 10 percent more snacks, and a 5 percent rise in body mass index … British psychologist Dr Aric Sigman, who has reviewed 30 scientific papers on TV and computer-screen viewing, said that governmental 'advice on TV is conspicuous by its absence. Politicians are scared to take on the entertainment industry, because that industry also provides political news.'"[289]

Based on Nielsen and other studies, the average American spends between 25,000 and 60,000 hours passively watching TV by their early 20s[290]—so their minds are less developed by that number of hours. This deeply compromises our faculties of reasoning. Adults who grew up watching average amounts of TV, including educators and scientists, have suffered the same damage to brain function, mental health, and emotional development. This is the unhealthy condition most of us suffer today,[291] and our schools reflect that fact.

Make Everything Easy (The New Normal)

When students fail in a particular requirement, one way to respond is to push students to work harder and succeed. Another way is to just get rid of the requirement, so students can graduate without learning or growing. Part of the new normal is to make everything easy (until later life when students have to compete in the workplace).

Make Everything Easy (Algebra)

In the new normal, we adapt by accepting our deterioration, for example, in programs like a San Francisco Middle School approach to simply remove Algebra 1 from the curriculum.[292]

One parent of an 11-year-old student "admits he has high expectations for his children. He also has high expectations for San Francisco Unified, which is why he and many parents like him were outraged when they learned Algebra 1 will no longer be taught in middle school under Common Core, the state's new academic standards."

His 11-year-old daughter says, "delaying Algebra 1 is going to hurt gifted students because some classes are 'too easy' or 'aren't very challenging' for high-achieving students."

"A growing group of San Francisco parents...are now pressuring the district to create more options so they don't have to choose between private school or paying extra for advanced math classes." As one of the parents whose 13-year-old son will be in eighth grade explained, "'It's very disappointing to me that our education system is really starting to be this cookie-cutter approach...It's not feasible. ... I really don't want [my son] to lose his engagement in school because it's not moving at a fast enough pace for him."

Make Everything Easy (Math and Reading)

In a similar make-it-easy move in Oregon, a 2021 bill ended both math and reading proficiency requirements for high school graduation.[293]

The governor "did not hold a public signing or issue a press release regarding the passing of Senate Bill 744 on July 14, and the measure...was not added into the state's legislative database until more than two weeks later on July 29, an unusually quiet approach to enacting legislation." One House member said "I worry that by adopting this bill, we're giving up on our kids." Otherwise, the bill quietly took effect without much comment.

Make Everything Easy (Homework, Deadlines, and Attendance)

In another example of the new normal, many public schools across the US are adopting a new program that lowers educational standards in several areas. The program says there will be no more deadlines for assigned work; homework is downplayed (meaning, less of it, and homework counts less towards the final grade); and no penalty for missing class.[294]

This move strikes at the heart of our educational system's erstwhile reason for existence, which was to develop strong study habits, organizational skills, time management, reliability, intelligence, and discipline (in addition to teaching math, reading, and writing). The program undermines all of those basic areas of development. Education's worst enemy could not have devised a more insidious scheme to enfeeble our youth and incapacitate the adults of tomorrow.

"The Clark County School District (Nevada)—the nation's fifth-largest school system—has joined dozens of districts in California, Iowa, Virginia and other states" in adopting the new program.

One teacher explains, "Homework is typically played down and students are given multiple opportunities to complete tests and assignments." The teacher added, "'They're relying on children having intrinsic motivation, and that is the furthest thing from the truth for this age group.'" In addition, "students are given multiple opportunities to complete tests and assignments." As a result, some don't turn in homework at all, knowing they could do it later. "Lessons drag on now, because students can turn in work until right before (final) grades are due."

As is obvious to many teachers, and to everyone else except the public school decision-makers, the above make-it-easy programs may temporarily (and artificially) bolster graduation rates and GPAs, but they will hurt students in the long run. In this new normal, grades and graduation no longer signify achievement—it's more like a birth certificate than a diploma.

One high school senior in the new program observed that the program provides incentives for poor work habits. He also noted that since absenteeism no longer counts against grades, "even classmates in honors and Advanced Placement classes are prone to skip class now unless there is an exam." That seems natural enough. If 16- and 17-year-old students are not required to attend class, why would they? "There's an apathy that pervades the entire classroom," he said.

This is not an isolated school or even school district. This program is growing more common and has spread to dozens of districts in Nevada, California, Iowa, Virginia and other states. "'We're really setting students up for a false sense of reality,'" one of the teachers warned.

"Some teachers and students say the changes have led to gaming the system and a lack of accountability." But are they really "gaming" the system? Gaming implies finding a clever loophole to circumvent the rules. But that is not what these students are doing. This is a straightforward adaption to the rules as given. Students are simply adapting to the new reduced requirements and lack of rules. Lower expectations naturally leads to lower performance—lack of rules leads to unruly behavior. Students are rationally and sensibly doing what is required–no more, no less. They are following the lead of the school system. The school is promoting slack behavior. There is not much of a system left to "game."

Just as it is the student's rôle is to perform according to the teacher's requirements—the teacher's rôle is to lead their students in a way that develops strong study habits, organizational skills, time management, reliability, intelligence, and discipline. These are necessary to succeed as an adult. Unfortunately, the new program discourages those very skills.

The teacher's rôle is also to prepare students for real-world jobs where a single missed day or one missed deadline could mean someone else gets the next promotion or raise (and the loser in this competition may never even realize why they were passed over). The new program is putting students last and setting them up for frustration and failure.

As dozens of studies that we have already cited in this book show, the "make-it-easy" trend follows predictably from our society's screen immersion and screen dependency. But for those who remain high performers, making school easier suffocates and punishes them for their high performance, and threatens to drag them down in the process.

In the above example, the Clark County school district faces challenges, "Clark County, NV public schools have an average math proficiency score of 24%...and reading proficiency score of 42%".[295] It is certainly a deep hole to climb out of. But as noted earlier, lagging performance presents a choice: push students to work harder and succeed; or just get rid of the requirements, so students can graduate without learning or growing. We are seeing our new normal of underachievement play out in the choices we are making.

Coronavirus Pandemic: Our Response Affirms the New Normal

Our decades-long declines in student learning and student performance, combined with the Coronavirus pandemic, created the perfect storm as educational outcomes rapidly deteriorated from bad to worse. A study published in August 2022 looking at more than 3 million children in more than 150,000 classrooms found that below-grade-level teaching has become the new norm.[296]

"'Students are spending even more time on below grade-level work than they were before the pandemic,' the report said. 'Students on the ReadWorks platform spent about a third of their time engaging with below-grade-level texts and question sets. In fact, they received 5 percentage points more below-grade-level content' than before the 2021–2022 school year."

The broad-based nationwide study also shows "that students were just as successful on grade-level work as they were on below-grade-level work. So their teachers rushed to give them more grade-level assignments, right?"

Wrong. Despite finding that students could handle harder requirements and grade-level instruction, teachers still opted to give students below-grade-level instruction, because it is easier. "Hundreds of teachers and much data over many years have convinced me that too many schools think the best way to educate kids is give them easy stuff…Instruction in the 2021–2022 school year suggests this bias in favor of dumbing down instruction is still with us."

Economically disadvantaged students get the worst of it, as "'Students in schools serving the most students in poverty spent about 65% more time on below-grade-level texts and question sets than their peers in the most affluent schools,' the report said."[297]

This trend is especially concerning in light of the fact that "Our past research, focused on mathematics, has shown that when students fall behind, providing access to grade-level work … is the best way to help them catch up—and that delaying access to grade-level work practically guarantees they will fall even farther behind."[298]

Our pervasive spread of below-grade-level teaching, combined with our pervasive spread of "screen learning," make recovery from the latest setbacks less likely than ever. We had a choice on how to respond to the post-pandemic challenge. Again we see our new attitude of surrender to the "It's Too Hard" mentality, and to "Just Make It Easy."

Coronavirus Pandemic: The Screen Learning Aftermath

The educational collapse illustrated in the previous section was bound to show up later in the young workforce. This section looks at some manifestations.

An August 2023 *Wall Street Journal* article reveals that screen learning is "gumming up workplaces around the country." The article elaborates, "It is one reason professional service jobs are going unfilled and goods aren't making it to market. It also helps explain why national productivity has fallen for the past five quarters, the longest contraction since at least 1948, according to the U.S. Labor Department."[299]

The shortcomings of the screen-taught generation run the gamut from general knowledge to specific skillsets. Employers are lowering expectations when they hire. "Then they are spending millions to fix new employees' lack of basic skills." Some additional examples of the screen-learning aftermath include:

- Pass rates on national certifications and assessment exams taken by engineers, office workers, soldiers and nurses have all fallen.
- On national standardized tests, the scores of fourth- and eighth-graders fell to 30-year lows.
- Scores for college admissions exams dropped.
- A notable decline in the younger workers' scores among about 10 million assessments a year that evaluate prospective employees. The skills test used in hiring measures reading comprehension, verbal-and-communication skills, grammar, spelling, numeracy, and attention to detail.
- Sharp declines in the behavior of job applicants as well as their performance on employment exams.
- Younger workers lack basic soft skills such as being able to work with other people, and tend to be more easily stressed and easily frustrated.

The falling scores among prospective engineers in the Fundamentals of Engineering exam "means fewer engineers on the job and a lower degree of competency among those who make it," according to David Cox, CEO of the National Council of Examiners for Engineering and Surveying. "The sharpest declines in scores came on questions measuring the most specialized knowledge. Structural engineers failed to answer questions about the use of trusses in the construction of bridges and roadways … These are areas that are very much involved in public safety," according to Cox.[300] These are the engineers who will design the nation's infrastructure in the years ahead.

From engineering to nursing: During the pandemic more than 100,000 nurses left the field. One difficulty in filling the gap is that up-and-coming students are failing out, or if passing the program, failing the post-graduation certification exam.

"More students who do enroll struggle to earn passing grades, said Patty Knecht, Vice President of Ascend Learning Healthcare, a private company which helps train medical professionals. And even if they do graduate, more are struggling to pass a certification exam. By then, they may already be on the payroll but unable to work. The delays cost hospitals an average of $42,000 per student who fails the certification exam, said Knecht."[301]

As one nursing school student admits, "When I got here I realized I wasn't ready for nursing school. I realized I didn't know how to study."

In other areas such as retail and customer service, lack of development takes other forms. Jerrica Moses, national recruitment manager for Senture, a Kentucky–based company, says workers under 25 cannot handle stress and are easily frustrated. Others in the research article point out that "young employees haven't been held accountable for things like finishing homework assignments, (which) has led to a decline in motivation. The younger workers are not even trying to be productive. "If they're not told what to do, if someone isn't managing every second and keeping them busy, their inclination is … to do nothing." Another supervisor said she is frequently surprised by the younger employees' lack of motivation.[302]

The educational pipeline does not bode well for our future. "Janet Godwin, chief executive of ACT, the nonprofit organization which administers the college admission test of the same name, said more high-school graduates today lack the fundamental academic skills needed for college and the workplace, with low-performing students facing the steepest declines."[303]

To accommodate the declines, "many college professors restructured curricula for students who lack basic study skills." For example, University of Alabama professor Mike Altman says he has "narrowed his curriculum to give his students more time to master basics" due to students' lack of reading, writing and critical-thinking skills.[304]

The Coronavirus Pandemic brought into sharp focus our developmental failures in academic, technical, and professional arenas. It highlighted our deficiencies everywhere that requires initiative, intelligence, or a skill. When we vaguely blame the pandemic for these shortfalls, we should keep in mind that the pandemic was merely an indirect cause. The direct cause of the declines was the shift to universal screen learning.

The next few sections dig deeper into the screentime-based educational deterioration that was already well underway before the pandemic.

An 'A' for Everyone

Another aspect of our "Make Everything Easy" new normal is that A-Students are being artificially propped up by easy grading. Students are being fooled into thinking they are exceptional by the sharp increase in A's being handed out in schools, K–12, public or private, and in Ivy League and State Colleges.

Making school easier to accommodate lower-performing students didn't start with the COVID-19 pandemic. It didn't start in the 2000s, or the 1990s. The first sharp increase in handing out A's happened from the late 1960s to the early 1970s—exactly when the first generation of kids who grew up with TV entered the educational system.[305]

This is one of dozens of similar TV-effects that surfaced at that same time, for the same reason. It was the first generation that was "made stupid" by too much TV.[306]

Disproportionate awarding of A's has continued to rise steadily since the 1970s. And as Catherine Rampell's *Washington Post* article points out:

- "A's—once reserved for recognizing excellence and distinction—are today the most commonly awarded grades in America.
- "That's true at both Ivy League institutions and community colleges, at huge flagship publics and tiny liberal arts schools, and in English, ethnic studies and engineering departments alike. Across the country, wherever and whatever they study, mediocre students are increasingly likely to receive supposedly superlative grades.
- "Analyzing 70 years of transcript records from more than 400 schools, the researchers found that the share of A grades has tripled, from just 15 percent of grades in 1940 to 45 percent in 2013"[307]

It's not because students are "doing better than ever." Students in the US are doing worse than ever. As Ray Williams' *Psychology Today* article notes: "After leading the world for decades in 25-34 year olds with university degrees, the U.S. is now in 12th place."[308] While we fool ourselves with more A's than ever, with absurd GPAs like 4.9 on a 4.0 scale, the rest of the world is leaving us in the dust.

More warning signs of our doing worse than ever come from a recent Educational Testing Service (ETS) report: "Indeed, while millennials are often portrayed in the media as being on track to be our best educated generation ever, their skill levels are comparatively weak." As compared with the same age group in other countries, "the comparatively low skill level of U.S. millennials is likely to test our international competitiveness over the coming decades. If our future rests in part on the skills of this cohort—as these individuals represent the workforce, parents, educators, and our political bedrock—then that future looks bleak."[309]

As we might expect from these findings, a *Scientific Learning* report notes that "Watching more TV in childhood increases the chances of dropping out of school and decreased chances of getting a college degree, even after controlling for confounding [socio-economic, etc.] factors … Children who watch too much television are likely to:

- "Forego fantasy play, which is critical to brain development because it helps kids understand symbolism, the foundation of reading.
- "Fail to question and develop alternative understanding and explanations which leads to creative problem-solving.
- "Develop weaker language skills.[310]

As psychologist Dr. Jean Twenge points out, "8th-, 10th-, and 12th-graders in the 2010s actually spent less time on homework than Gen X teens did in the early 1990s."[311] Each generation is getting worse.

The warning signs of our descent continue as the *New York Times* reports from the Organization for Economic Cooperation and Development, the US scores 17th in numeracy testing, among 16- to 29-year-olds with a bachelor's degree. "While results vary somewhat depending on the subject and grade level, America never looks very good. The same is true of other international tests. In PISA's math test, the United States battles it out for last place among developed countries, along with Hungary and Lithuania."[312]

Professor Tom Nichols points out the detriment to students: "Colleges and universities also mislead their students about their own competence through grade inflation. Collapsing standards so that schoolwork doesn't interfere with the fun of going to college is one way to ensure a happy student body and relieve the faculty of the pressure of actually failing anyone."[313]

As passing out A's like candy increases, reward becomes more disconnected from performance. Students are trained to feel entitled to an A grade as a matter of course. "Professors marvel at the way students now shamelessly demand to be given good grades, regardless of their work ethic, but that's exactly what you would expect if the student views themselves as a consumer, and views the product as a credential, rather than an education."[314]

If students' self-esteem is temporarily propped up by fraudulent GPAs, it will be mercilessly torn down by real-life competition later. That's not doing students a favor, it's a cruel trick being played on them. Some rationalize that exaggerated grades nurture students' self-esteem.

Instead, as Rampell notes, we are "hampering the ability of students to compete in the global marketplace."[315]

To be fair, many teachers are under pressure to give inflated grades. That trail leads back to money, with funding based on school performance (such as average GPAs and graduation rates), whereas true student development is not part of the funding equation.

Administrators are under similar pressure to put pressure on teachers, for the same reasons. In some cases their livelihoods and careers are at stake: deliver high grades to get more funding. Students are the losers. When we put our kids' learning into the hands of bureaucrats competing for money, this is what we can expect.

At the college level, there are even more politics at play. Graduates with higher GPAs and more honors are more likely to get job placement, regardless of the lack of real learning. That means colleges can say more of their students get jobs in their field of study, which attracts more new students, which increases revenue. It's a business. As Nichols warns, the commoditization of degrees, and the insatiable demand to keep printing more of them, at the overstressed assembly lines of diploma mills in America, we're descending into a destructive spiral of credential inflation.[316]

Whether K-12, public or private; or college, Ivy League or State School; student performance is the least significant factor in the grades they receive, and the least significant factor in our educational policies.

Educational deterioration and commoditization traces back to greed, negligence, immediate gratification, artificial self-esteem, and lower functioning minds enacting policies that work against intellectual development. This is evident in government, in colleges, in public schools.

Corruption is nothing new. But what is new, is the massive scale of this fraud beginning in the 1960s and 1970s, which never happened before in history, and which continues to get worse today.

Other countries are starting to imitate American poor performance. For example, in Britain: "An international survey by the Organisation for Economic Cooperation and Development has found standards of literacy and numeracy among school-leavers in England and Northern Ireland to be among the lowest in the developed world. Shockingly, older people leaving the workforce were better educated than those joining it. This may be the first time in recorded history that such a phenomenon has occurred, with the young worse educated than their parents."[317]

Declines in higher levels of intelligence have also started throughout Europe, as a result of increased use of technology. As noted in a 2017 European study, "Intelligence levels are falling in advanced industrial countries across Europe."[318]

As other countries drop to US levels of screen-related lower functioning, we will witness the spread of educational deterioration globally.

Mental Comfort versus Mental Development

By exaggerated nurturing and over-protectiveness, we send the message that mental comfort is more important than mental development. But why can't we have both? We can't have both because mental comfort and mental development are mutually exclusive. Learning requires discomfort—if it is not painful, it is not learning.

Instead of pushing students to greater accomplishments, we train students to resent and resist hard work. We reward them for failing to reach beyond their comfort zone. At the first sign of mental discomfort, children rebel and act defiant, because we conditioned them to. It's an unnatural response embedded in the child's brain that stems from too much screentime. As a result, basic mental development requires a level of mental discomfort that is beyond students' capacity to tolerate. This deterioration of learning-pain-tolerance spreads throughout the educational process, K-through-College-through-Career.

As *The Death of Expertise* author Professor Tom Nichols warns, "the protective, swaddling environment of the modern university infantilizes students and thus dissolves their ability to conduct a logical and informed argument. When feelings matter more than rationality or facts, education is a doomed enterprise."[319]

We want to "stretch our boundaries" and "reach outside our comfort zone," and so we should. But it requires pushing students to learn that lesson. We have failed in this fundamental educational standard. Despite the constant drum beat of complaints regarding our educational system, it keeps getting worse.

As the preponderance of scientific evidence makes clear, being raised in the US today is fast-becoming a developmental death sentence. There is indication that universities may be dredging the bottom of the GRE scorers in order to find American-born graduate students, and yet still must take mostly the foreign-born high-scorers in the sciences.[320]

A "new report provides a breakdown on international enrollments by discipline and institution, showing that there are graduate STEM programs in which more than 90 percent of students are from outside the U.S."[321]

With the more recent mental health issues now known to be associated with smartphones, we should beware of the profound developmental impacts that have long been associated with television as well. As noted in the Reuters report *Screen time linked to ADHD symptoms in teens*, "Older forms of screen time—like television watching and playing video games on consoles—have long been linked to an increased risk of ADHD and other emotional and behavioral problems."[322]

Attention span, focus, and concentration issues follow as screentime makes the brain more passive. Screentime conditions us to crave passive comfort. It systematically undermines our ability to grow and learn, because it reduces our tolerance for the discomfort required to grow and learn.

A Simple Process

As *The Motivation Myth* author Jeff Haden aptly points out, "Improving feels good. Improving breeds confidence…so you naturally want to keep improving."[323] One of the best lessons in life is that "my aversion to 'hard' goes away once I break a sweat."[324] When "hard" is not required, the lesson is never learned.

Teachers who create demanding lesson plans, explain subjects, assign work, test progress, and grade hard, produce the most successful graduates. Students who listen quietly and attentively, do their homework, study for tests, and perform with excellence go on to live the most successful lives.

This simple process sounds farfetched today, even though it was done precisely this way, with consistent success, for hundreds of millions of students, for hundreds of years, until the latter twentieth century, when TV began lowering mental functions and educational expectations. Today, healthy educational experience is impossible with our post-2010 collapse of mental health and brain function from virtually universal screen addiction. The main truth learned is that learning is not part of the educational program.

Homeschooling

Hope for tomorrow comes from involved parents, who supplement school with home education (or just homeschool altogether), who stop letting kids watch TV, and minimize any kind of screen exposure on any device.

For those who still question the efficacy of homeschooling, here are the results from a recent study:

"The study included almost 12,000 home-school students from all 50 states who took three well-known standardized achievement tests...The students were drawn from 15 independent testing services, making it the most comprehensive home-school academic study to date.

- "In reading, the average home-schooler scored at the 89th percentile; language, 84th percentile; math, 84th percentile; science, 86th percentile; and social studies, 84th percentile. In the core studies (reading, language and math), the average home-schooler scored at the 88th percentile.
- "The average public school student taking these standardized tests scored at the 50th percentile in each subject area.
- "The average home-school test results continue to be 30-plus percentile points higher than their public school counterparts.
- "In a sentence, home-schooling is a recipe for academic success."[325]

Most people would like to see public education be the best option to prepare kids for the real world. But at present public education is the worst option.

A College Eye View

From this author's firsthand experience both as a graduate student and as a university administrator, colleges are requiring less and less of their students. As *The Death of Expertise* author Professor Tom Nichols noted, "In the worst cases, degrees affirm neither education nor training, but attendance. At the barest minimum, they certify only the timely payment of tuition," and adds that "students now graduate believing they know a lot more than they actually do," while "Intellectual discipline and maturation have fallen by the wayside."[326]

Colleges and Universities are places to make special memories of youthful carefree years, but not necessarily places of rigorous development of any kind.

Between two students with the same degree from the same college, with the same GPA, there might be one with very little learning, and another with a lot of learning. Requirements for assignments, and standards for "passing a class," are a lot lower than in the past. Required reading has become less challenging. There is a steady and systematic reduction of difficulty. As Nichols observes, "Less is demanded of students now than even a few decades ago. There is less homework, shorter trimester and quarter systems, and technological innovations that make going to college more fun but less rigorous. When college is a business, you can't flunk the customers."[327]

As a result, the lesser-learning student can easily graduate with the same degree as the student who learns a lot. Nichols echoes this new fact of college education today: "'College graduate' today means a lot of things. Unfortunately, 'a person of demonstrated educational achievement' is not always one of them."[328] Graduates not only fail to achieve expertise in anything, they even don't learn enough to *recognize* expertise in others.[329]

For those students who want to learn more, who appreciate the sacrifice and the rewards of painful effort and fulfillment, who know learning has nothing to do with entertainment, who push themselves to learn on their own, the university can provide guidance and resources.

Universities still have dedicated mentors, old-school professors, just waiting for a few students who want something more from college. Those professors will devote extra time to work with those students, to make the diploma stand for something more substantial. But that kind of accomplishment is no longer required in order to receive a diploma. It doesn't make the piece of paper, it just makes the piece of paper meaningful.

There is some good news: Some colleges are changing admissions tests to filter out students with spoiled attitudes, to get a higher proportion of more driven students. Psychological testing showing an attitude of "I've got to put in more effort" instead of "It's the professor's fault I'm not learning" might open more doors to get into college ahead of others with a higher GPA from high school.[330]

Multitasking

As new behaviors grow more common in society, we have to create a label for them. As screentime systematically impairs our ability to concentrate, and reduces our attention spans, the behavioral symptoms of this new breakdown need a label. The label we have given to a common practice emerging from impaired concentration and reduced attention span is "multitasking."

Contrary to conventional wisdom, multitasking is not the friend of productivity. Quite the contrary, productivity has no worse enemy.

As *Emotional Intelligence 2.0* author and clinical psychologist Travis Bradberry notes, "Ultra-productive people know that multitasking is a real productivity killer. Research conducted at Stanford University confirms that multitasking is less productive than doing a single thing at a time. The researchers found that people who are regularly bombarded with several streams of electronic information cannot pay attention, recall information or switch from one job to another as well as those who complete one task at a time."[331]

Multitasking is a concept that arose from digital lifestyles where we can easily switch between screen tasks. It is a behavior pattern derived from the passive screen-culture of hectic-but-ineffectual movement that passes for activity. The loss of concentration and attention span compromises our effectiveness when dealing with a multitude of tasks. What people call "multitasking" (e.g., having several assignments going at the same time, rapidly switching between tasks) makes people less effective, wastes time, and reduces the quality of everything we do. The mind functions at a lower level when multitasking, making it capable of only simplistic activity, and incapable of complex work.

As Doctors Cynthia Kubu and Andre Machado explain in *The Science Is Clear: Why Multitasking Doesn't Work*, "Empirical research has demonstrated that multitasking with technology … negatively impacts studying, doing homework, learning and grades," adding that "the more we multitask, the less we are able to accomplish."[332]

As Dr. Bradberry further explains, "Multitasking actually sacrifices your quality of work, as the brain is simply incapable of performing at a high level in multiple activities at once."[333]

No matter how young and energetic, no matter how old and experienced, multitasking yields inefficiency, missed deadlines, and elevated stress. Multitasking ultimately takes longer, makes us less effective, and lowers the quality of everything we do. It has become an excuse for carelessness. As several studies have demonstrated, multitasking is a fragmented, shallow mode of limited surface engagement, symptomatic of an easily distracted brain.[334]

Focusing on big chunks of work requires deeper concentration and longer attention spans—two abilities shown to be undermined by screentime. The solution is to reduce screentime, and where possible, dedicate deeper prolonged attention to one big chunk of work at a time. It will reduce stress and headaches. It will improve quality and productivity in the short run and in the long run.

Handwriting versus Typewriting

Computers can be used for constructive work and creative purposes, whereas watching videos is a purely passive experience. Having said that, even when we think a computer is helping us learn, it might actually be slowing us down.

A study using undergraduates from UCLA and Princeton, showed students taking notes on a laptop score significantly lower than students who took notes by hand. When the test was a half-hour after the note taking, they scored about the same in the fact-based questions, but laptop students score poorly on conceptual questions, while handwritten note takers scored well.

Testing a week later, laptop students scored poorly on both fact-based and conceptual questions, while handwritten note takers scored well in both areas.[335]

Using an EEG to track brainwave activity yields the same results. A study conducted by Professor Audrey van der Meer and others at the Norwegian University of Science and Technology collected 500 data points per second during forty-five-minute tests for each participant in the study. "The results showed that the brain in both young adults and children is much more active when writing by hand than when typing on a keyboard."

"The use of pen and paper gives the brain more 'hooks' to hang your memories on. Writing by hand creates much more activity in the sensorimotor parts of the brain. A lot of senses are activated by pressing the pen on paper, seeing the letters you write and hearing the sound you make while writing. These sense experiences create contact between different parts of the brain and open the brain up for learning. We both learn better and remember better," Van der Meer said.[336]

In a separate study, van der Meer notes that "when we measure the brain activity of people who write by hand, we see that they form more connections in the brain than when they write using a computer." She adds, "When people write by hand, we see noticeably more nerve activity in the parts of the brain that deal with memory and interpretation of new information. This activity plays a key role in the learning process." Study after study shows that students learn new information best when they take notes by hand. The big takeaway here is that "Children must first learn to write by hand, starting from Year 1 of school, so that they can form the neural networks that create the best possible foundation for learning." It's no wonder that "this year (2024), 20 states in the United States reintroduced handwriting at school."[337]

Other studies, reported by Dr. Perri Klass, professor of journalism and pediatrics at New York University, support the importance of handwriting over keyboard writing for children.[338] In handwriting, "You have to see letters in 'the mind's eye' in order to produce them on the page," said Dr. Virginia Berninger, professor of educational psychology at the University of Washington and the lead author on the study. It's a critical region of brain development, where the visual and language come together, she said. Typing the letters doesn't generate the same brain activation, added Karin James, professor of psychological and brain sciences at Indiana University. Without handwriting, children's brains don't distinguish letters; they respond to letters the same as to a triangle, she said. Conversely, writing by hand improves brain development and brain function. The conclusions were further supported by other studies that show college students who are writing on a keyboard understand and remember less than if they were writing it by hand, added Dr. Laura Dinehart, associate professor of early childhood education at Florida International University.[339]

Handwriting improves brain development and brain health for children. Note taking by hand improves understanding and retention for all ages. Even when screens are used actively as a learning tool, it turns out, non-screen methods work better.

Reading comprehension follows the same rule. A recent study compared screen-device reading versus reading a physical book. Students read a short story. "Half the students read on a Kindle, the other half in paperback. Results indicated that students who read on print were superior in their comprehension to screen-reading peers, particularly in their ability to sequence detail and reconstruct the plot in chronological order."[340]

Several studies at the University of Maryland, reported by Professor Patricia A. Alexander, yielded similar results. The screen experience actually fooled the students into thinking they studied better on screens. "Students judged their comprehension as better online than in print … Paradoxically, overall comprehension was better for print versus digital reading … when it came to specific questions, comprehension was significantly better when participants read printed texts."[341]

Other academic studies and educational research reaffirm the fact that reading a physical printed book or article leads to greater comprehension, deeper concentration, longer attention, and improved memory of the text, over reading the same text on a digital device.[342]

Screens not only degrade student comprehension, they give students false confidence that they comprehend better with screens. This is a familiar phenomenon, as we saw in the Dunning-Kruger Effect section earlier in this book: lower comprehension equals higher confidence in being right. It seems there is no end to the ways screens undermine our development and our competence.

Reading Good Books Enhances and Lengthens Life

Great works of fiction are those with a more layered, complex investigation into the human condition, written in an artistic language resulting from painstaking development, so that it appears effortless.

Reading is a demanding process that yields deeper pleasure and more meaningful experience than movies and video. Every good book we read creates more depth in our thinking.

We also gain a deeper understanding of other people: "an influential study published in *Science* found that reading literary fiction (rather than popular fiction or literary nonfiction) improved participants' results on tests that measured social perception and empathy, which are crucial to 'theory of mind': the ability to guess with accuracy what another human being might be thinking or feeling."[343]

It turns out we can enjoy our books longer as well, because reading them makes us live longer, according to another study: "Overall, the researchers calculated that book reading was associated with an extra 23 months of survival. … Reading magazines or newspapers didn't have the same effect ... it's the deep engagement required by the narrative and characters of fiction, and the length of both fiction and nonfiction books, that increases cognitive skills and therefore extends lives."[344]

In addition to empathy and the deeper cognitive health that we gain from great literature, it also offers useful perspectives in life. In his *Wall Street Journal* piece "I'm Revisiting the Books of My Youth,"[345] columnist Danny Heitman remembers that when he first read *The Norton Anthology of American Literature* in 1982, it left him cold. "I wanted to be a newspaperman, and the musings of Henry David Thoreau, Ralph Waldo Emerson and Emily Dickinson struck me as far removed from real life. I dutifully plowed through the assigned readings and churned out the term papers, relieved at the end of the semester when I could put the rarified writings of a distant era behind me."

Heitman goes on to say, "I'm a much better reader of the classics today than I ever could have been at 18…As the obligations of marriage and parenthood kept me home more often, I reread passages from Thoreau's *Walden* for instruction in how to savor small moments outside my doorstep. In the wake of family deaths, I found Emerson's quiet resolve after his own losses an inspiration. Dickinson, whose poems remained open to joy as the country careened toward the Civil War, offers me a model in seeking serenity amid social division."

"'The trouble with education,' Margaret Ayer Barnes observed in 1930, 'is that we always read everything when we're too young to know what it means. And the trouble with life is that we're always too busy to reread it later.'"

"'Her remark rings true,' Heitman says, 'I'm trying to make more time to reconnect with authors I met in my youth. My old Norton, worn by the years but still intact, seems destined to see me through.'" This example reminds us of why the greats are so precious, and so useful.

Reading lengthens life (and improves the quality of it); watching TV shortens life[346] (and ruins the quality of life). What a choice.

How Our Presidents Represent Us

The steadily decreasing aptitude of the US population is reflected in the declining literacy of presidential speeches. It provides yet another reference point that reflects the developmental decline of Americans and correlating it to the TV timeline.

Tracking trends over the course of 600 presidential speeches, starting with George Washington, up to the second term of Barak Obama, the trend shows a steady decline.[347] One attempt to explain it is to say presidents spoke to voters, and in the early 1800s, only well-off landowners could vote. With a broader voting public, the argument says, literacy of speeches had to be lowered to communicate. But that argument doesn't explain it.

Education has never been more broadly available to everyone, K–12, than it is today. How people use their educational opportunity, and the quality of institutions of education, are what make the difference.

In 1800, even landowners were often "uneducated" in the sense that they didn't get much formal schooling. Yet they possessed much higher reading levels and aptitude than today's graduates. Self-taught reading-and-learning lifestyles were common. For example, George Washington had a "meager education"—he never received more than the equivalent of an elementary school education.[348]

George Washington had a meager elementary school education and was not highly literate by eighteenth and nineteenth century standards, but his speeches were at today's Ph.D. reading level.

Twenty-first century speeches average about 9th grade.

There is a steady record of graduate-level presidential speeches in the eighteenth and nineteenth centuries. The only time someone like Thomas Jefferson spoke below today's college level was speaking to people who knew English only as a second language, such as his 1806 speech to the Cherokee Nation (8th-grade level), or his 1803 speech to the Choctaw Nation (10th-grade level). George Washington spoke to the Cherokee and Seneca Nations closer to 11th-grade level.[349]

People with no school whatsoever, who knew English only as a second language, had equal or higher reading levels than today's mass educated culture.

Speeches to English-speaking American citizens remained at today's college level to Ph.D. level until the twentieth century. In the past, the general public made themselves more literate than people today, with much less formal schooling than today. Adults taught children or they learned on their own.

To achieve such levels of learning without formal education, they must have appreciated delayed gratification, long-term rewards of painful study. No one forced them to go to school and sit through class. They wanted the pain, because they understood the gain. They had the initiative on their own, and they cherished higher functioning as an important value.

The real drop in presidential-speech reading levels begins in the 1920s as radio starts to replace reading as an information source. Public literacy drops more dramatically and rapidly after TV hit the scene. It is reflected in the drop in presidential-speech literacy to 9th-grade levels by the late twentieth century, where they remain today.[350]

Self-Mocking

Instead of valuing complex learning in all areas of the mind, our culture has sunk into a kind of surrender, celebrating getting dumber and dumber. The decline and the surrender show up in social media tropes, on T-shirts, in ads, saying things like "Poetry is hard, Mmmm Bacon" or "Math is Hard, Let's go Shopping." Sure, these are meant to be funny. And people say, "It's just a joke." Yet it accurately reflects today's mindset and skillset. It expresses our true cultural values. It mirrors our true response to difficulty. Jokes are supposed to employ exaggeration in order to be funny. The joke is no longer an exaggeration, but we are still laughing.

Far from being funny, it is sickening to see the population self-administering brain deterioration, surrendering, and laughing at the dreadful process.

Everyone chuckles and says, "Hey, I'm no good at math," with a wink that says, "Nobody cares anyway." Like the example given in the article "We're evil, and proud of it," TV ads mock the idea of TV melting our brains even though the actual deterioration of the brain is happening while advertisers and viewers laugh at the joke together.[351]

CHAPTER 6: ARRESTED DEVELOPMENT

In this chapter we explore several ways children in the Screen Age take longer to develop than they did in prescreen eras.

Overparenting

As screen-based lifestyles separate and isolate parents from children, parents become less attuned to issues in the life of their child. They miss opportunities to help, opportunities that would have fostered natural parent-child bonding.

Screen-related parent-child estrangement has become a common issue, beginning in the mid-twentieth century and worsening every generation since. Alienation worsened dramatically post-2010 as smartphones reduced human contact even more.

As Child Psychiatrist Victoria Dunckley points out in *Psychology Today*, "research indicates that the more time a child spends using the Internet, the less healthy the parent-child relationship becomes."[352]

Screentime's correlation to parent-child alienation is not new, and in fact is a well-documented fact in today's screen-centric family dynamics. For example, a recent *Pediatrics & Adolescent Medicine* report shows that "More time spent television viewing and less time spent reading and doing homework were associated with low attachment to parents for both cohorts," and that, "more time spent playing on a computer was also associated with low attachment to parents." The study further noted that "more time spent television viewing was associated with low attachment to peers."[353]

iGen does not go out as much as any previous generation. But ironically, they spend even less time with other people in the house. San Diego State University Professor of Psychology Jean Twenge observed, "One of the ironies of iGen life is that despite spending far more time under the same roof as their parents, today's teens can hardly be said to be closer to their mothers and fathers than their predecessors were." They are expert at tuning out their parents so they can focus on her phone. "So what are they doing with all that time? They are on their phone, in their room, alone and often distressed."[354]

Often, parents try to compensate for the distance that they caused, by lavishing too much sympathy on their children's tiniest troubles.

Over-attention reveals being out of touch just as much as under-attention. Over-attention is an exaggerated attempt to "show they care" for a moment. It is forced and unnatural against the lack of bonding from screen isolation. Over-parenting leaves kids with escalated anxiety when faced with challenges in the real world.[355]

Parents also compensate for lack of quality time together by over-serving kids, especially while they are watching TV. It becomes a habit, and an expectation, and another developmental impediment that undermines the child's ability to grow up.[356]

Additionally, parents damage their children with over-praising minor accomplishments, with disproportional accolades. Their children grow up to expect exaggerated rewards. "While experts say it's natural for humans to seek attention, these young people revel in it. They're accustomed to being noticed, having been showered with awards and accolades."[357]

Overparenting is now continuing into kids' late teens and young adulthood. Today, parents take care of their kids' college applications, financial aid forms, bank accounts, job-search, and give them credit cards. Some parents even want to be present at their adult kids' job interviews (a humiliation that causes further estrangement). Meanwhile, overparenting undermines the young adult's coping skills, and further delays the young adults' ability to competently function in the real world.[358]

One of the symptoms of overparented children is a decline in the ability of motivation to do for themselves. As Dr. Twenge reports, "In earlier eras, kids worked in great numbers, eager to finance their freedom or prodded by their parents to learn the value of a dollar. But iGen teens aren't working (or managing their own money) as much. In the late 1970s, 77 percent of high-school seniors worked for pay during the school year; by the mid-2010s, only 55 percent did. The number of eighth-graders who work for pay has been cut in half." Dr. Twenge adds that the onset of adolescence occurs many years later than ever before, pre-adolescent childhood stretches well into high school, with a drastically delayed ability to handle responsibilities of adulthood.[359]

Kids Grow Up So Slow These Days

Some obvious conclusions can be drawn from the studies and research we have discussed up to now. Self-management skills are a prerequisite to growing up and surviving in today's society. Growing up is a prerequisite to achieving independence and satisfaction in life. In the 1970s just as the legal adult age was lowered from 21 to 18, the functional adult age was rising from 21 to about age 30. It was at that time that parents starting saying things like "Kids grow up so fast these days!" Parents who didn't understand the impact of screentime were already becoming alienated from their children's reality. Children were growing up slower than ever by the 1970s and it has only gotten slower since then.

Because of TV, children became increasingly exposed to sex, drugs, violence, and corruption, on screen, all at the same time. Adults confused this kind of exposure with "growing up." Meanwhile, the brain in children and teens became less developed and more impaired every day. Most of the developmental activity that adds up to maturity was lost in their upbringing. TV exposure did not make children "grow up faster." It suppressed development in every area of the growing-up process.

As a result, children began staying in their parents' home to an older age, and that age has steadily become older.[360] Many reasons are suggested, such as cost of housing, marrying later, and others. These are surface symptoms of the deeper cause, which is simply slower development in every area of cognitive, social, and emotional maturity. Before screentime began thwarting this development, people were self-reliant, mature, responsible, started on their own, and started families, at much younger ages. The fact is, "Kids grow up so slow these days."

Adulting

The term "adulting"—to mean "acting responsibly" or "behaving maturely"—started appearing in social media around 2009. Usage of the term skyrocketed in 2016 (e.g., up 700% on Twitter).[361] That year, it was a runner up for Oxford's New Word of the Year.[362] It's an expression that arises organically from our new lifestyles of screen-dominated inactivity, and the resultant developmental shortfalls noted in the earlier section "Kids Grow Up So Slow These Days." As *dictionary.com* notes, "The concept behind *adulting* may stem from the fact that members of the rising millennial generation are going through major life stages (such as getting married, having children, buying a home, etc.) at much later ages than previous generations."[363] The *Urban Dictionary* provides the example, "Used by immature 20-somethings who are proud of themselves for paying a bill."[364] Former bureau chief for *TIME Magazine* cites similar examples, such as culture writer Madeleine Davies' observation:

> Madeleine Davies believes there is a 'self-congratulatory' vein to this, an expectation of millennials to be applauded for "fulfilling your basic responsibilities as a human," which she finds to be "pretty much the most childish thing imaginable."[365]

Adulting has come to mean something irritating and painful like we see in a child's recoil when they are told to eat their spinach. If it interrupts our binge watching eight hours of streaming shows, and jars us away from our immobilized stupor, even for one minute, we *hate* it. Adulting is a "necessary evil." Thus "acting responsibly" is a necessary evil. We have demoted constructive activity and emotional maturity from a virtue to a petty grievance with reality. Conversely, we now look favorably upon childish irresponsibility, craving more screentime, physically and mentally motionless and loving it.

The term adulting would be impossible to conceive of in prescreen society (hence, it was not conceived of in prescreen society). Only after the unnatural infantilization of the population could a word like adulting enter the language, even as a joke. But it is not a joke. It is a real complaint referring to those brief but unpleasant moments when we are forced out of our screen-induced lethargy to do something constructive. This new reality is the opposite of approximately 200,000 years of prescreen humanity whose lives were filled with constructive effort, when nonconstructive moments were a rare exception. Now constructive effort is the exception. We *adult* when we have to; we *infantile* when we can.

Goblin Mode

While the term adulting made runner up in the *Oxford University Press* 2016 Word of the Year, the term "goblin mode" won top honors as Word of the Year in 2022. "Although first seen on Twitter in 2009, goblin mode went viral on social media in February 2022."[366]

What is this beloved term, goblin mode? As described by Juliana Kim, fellow at *The New York Times* among other journalistic posts, "It's mindlessly binge-watching television without worrying about the time. It's eating snacks in bed without a care about leftover crumbs. And it's wearing the same pair of pajamas all week while working from home…a 'type of behavior which is unapologetically self-indulgent, lazy, slovenly, or greedy…'"[367]

We have seen these descriptors before, in the many studies describing the symptoms of screentime. The weakening effect of screentime to brain and body begets more screentime, and more terminology to describe its effects. Hence, adulting and goblin mode.

The same culture spawned both terms: the hated "adulting" (love to use the word but hate what it refers to) and the beloved "goblin mode" (love the word *and* love what it refers to). They are two sides of the same coin: a screen culture of negligence and incompetence. From this and many other similar trends we have seen so far, our society is not in a healthy place. Screentime is the root cause.

Interestingly, the 2022 3rd-place word was "#IStandWith,"—used on social media to stand up for a principle or a cause. Users fill in the name of the cause, but the term gained currency during the Russia–Ukraine war (i.e., #IStandWithUkraine). Approximately 300,000 people voted on which word should win the top prize. A small fraction voted for #IStandWith, while goblin mode garnered 93 percent of the vote. The lopsided dichotomy speaks volumes. But does Oxford's Word of the Year award mean anything significant? To many, perhaps not. Nevertheless, combined with hundreds of other cultural phenomena we have referenced so far, this reflection of our culture today is unmistakable.

Weakening Effect

We've seen a wealth of scientific data showing how screentime has suppressed human development. We see significantly increased carelessness, negligence, incompetence, and weakness, while screentime has undermined human abilities of all kinds across the population. This overall weakening effect has been largely uncommented in the mainstream for two reasons:

- First, the change happened gradually over several generations, and became the only reality we knew. It was gradual enough so that no one really noticed the extent of the loss.
- Second, the change happened to everyone, which made it even less noticeable. There was virtually no one immune to the deterioration, so no one against whom we could compare ourselves—no one with strong, well-developed, diverse abilities. The erosion of skill, talent, and strength was uniform and universal. Everyone deteriorated in lockstep, in gradual but steady decline. Without anything to contrast against, our loss of abilities was perceived as the normal way.

We experience increased frustrations, decreased initiative, less achievement, devalued virtues, and lack of satisfaction in life. If these are commented at all, they are attributed to all kinds of things other than screentime. We keep staring at the problem under the influence of deeper and deeper screen-induced lethargy. We give in to its weakening effects, without any notice of what's happening to us.

We used to build better opportunities for future generations, but we have changed that pattern. Because of our screen culture, we are losing the love of learning, development of skills, willingness to struggle, and values of hard work and responsibility. "What television is best at doing is transforming us into passive viewers disinclined to involve ourselves with improving the political, social and economic conditions in which we live."[368] Instead, we associate with lowest common denominators such as Reality Television for our standards of behavior. Digital networks become our closest personal connections.[369]

Attraction to the Difficult

Throughout history, people thrived on tough, painful mental and physical challenges. Healthy developed minds feel a strong attraction to difficult hurdles, pursuing them by nature with determination and enthusiasm.

During the mid- to late-twentieth century, when screens first began altering brain function, we began losing interest in the difficult. We were reshaped to crave whatever is easiest. Passive screen entertainment gave us this way of life, which has weakened us ever since. Instead of energy to meet challenges, we have numbness and remain passive.[370]

The fact alone that the average American watches 30 to 40 hours of TV per week, plus many more hours with other screen-device entertainment, is itself an attraction to the easy and effortless. That fact alone proves our switch from active and strong to passive and weak. No other activity or lack of activity or any other comparison or evidence is necessary to prove the point. This fact *by itself* proves that we lost active time and gained passive time.

This is not an indictment of any particular group of people or generation. Immerse any group of people into a sea of screens and you'll get the same result: passive dependency. Unfortunately for us, we are the ones who became so immersed. We are left with lower-functioning brain patterns and a culture of lethargy and atrophy. Removing TV and other video entertainment is the only way to fix it.[371]

One of the greatest gifts we could ever give to ourselves and to others, is to switch back to a natural, active lifestyle. Truer words were never spoken, than these by *The Motivation Myth* author Jeff Haden: "All of our success and growth comes from choosing the hardest and least comfortable way."[372]

Like most effort-related discomfort: "After the first few times you lean into your discomfort, you will quickly find that the discomfort isn't so bad, it doesn't ruin you, and it reaps rewards."[373]

PART II:
THE HIGH ROAD

CHAPTER 7: CHARACTER DEVELOPMENT

Throughout this book, and the many studies referenced, we have learned beyond any doubt that screentime contributes to mental passivity and deterioration. This chapter contrasts healthy development of character against unhealthy shortcuts to self-esteem.

Everyone desires a positive self-image, but excessive screentime frustrates that desire. Screenlife makes us more likely to look for easy shortcuts, to cheat our way towards exaggerated self-worth.

Bad decisions, for example, can undermine self-esteem. Deflecting responsibility for mistakes is a common self-esteem trick, often used when there is insecurity in this area. We make excuses for the consequences of our decisions. Two common ways of deflecting responsibility are

1. Inflated-ego: "I'm great and nothing was my fault. I did everything right. My problems were caused by others. Look at the bad luck I had to deal with, look at what happened that I couldn't control, look at what they did that made it harder to avoid, look everywhere else, but don't look at me."
2. Self-loathing: "I'm a no-good failure, there's no use trying. I'm such a mess, I can't do anything right. Why should I even care? I can't change who I am! I'm a failure and I might as well accept it."

Both are dishonest cover-ups for dodging ownership of our actions. It takes energy to change or improve. Failing such energy, we resort to the two dodges: "I'm great as I am, no change needed"; or, "I'm a total failure, no change possible, no use trying."

In both of these extreme-passive responses, there is no hope for fixing what's broken, because there's no accurate appraisal of the problem.

Even the "total-failure" case does not recognize what's broken accurately. It merely hides a fixable flaw behind an exaggerated self-loathing. As *The Motivation Myth* author Jeff Haden notes, "Hide from your weaknesses, and you'll always be weak. Accept your weaknesses and work to improve them, and you'll eventually be stronger—and more motivated to keep improving."[374]

The two dishonest tactics—inflated-ego and self-loathing—allow the passive avoidance of accountability for mistakes. Dr. Bradberry aptly sums it up: "If you are someone who thinks either *it's all my fault* or *it's all their fault* you are wrong most of the time."[375]

When we hide from responsibility for our mistakes and renounce ownership of our actions, we distort reality to protect our feelings—either with the inflated-ego tactic, or the self-loathing tactic. Unhealthy strategies such as these naturally occur in a life of screentime, because screentime displaces thousands of social situations and interactions that should have occurred in life, but didn't. Emotional development never had an opportunity.

The way to rise above these passive-avoidance tactics is to acknowledge responsibility and to keep improving: "Emotionally intelligent people know that success lies in their ability to rise in the face of failure."[376] By following the path of responsible accountability, we learn that "If I do good things, good things happen, if I do bad things, bad things happen, and if I do nothing, nothing happens."[377]

The result is amazing: better response, better results, better self-worth. "By holding yourself accountable, even when making excuses is an option, you show that you care about results more than your image or ego."[378]

The healthy response to mistakes is to accept our flaws and failings, while steadily improving. Blaming circumstances and other people surrenders your self-worth and your future to the control of others. As clinical psychologist Travis Bradberry explains, "Here's the worst thing about feeling sorry for yourself, other than it being annoying, of course: it shifts your locus of control outside yourself. Feeling sorry for yourself is, in essence, declaring that you're a helpless victim of circumstance. Emotionally intelligent people never feel sorry for themselves because that would mean giving up their power."[379]

Our choices create our journey and determine our destination. When mistakes happen, a genuine individual, in touch with reality, does not point at other people, or at circumstances, but looks in the mirror and says "I did this." The fault does not lay with society, parents, children, friends, job, or any other circumstance. This healthy understanding will never happen in front of a screen.

False Positive Self-Esteem

Our society's emphasis on positive self-esteem for its own sake conflicts with true self-worth. We can't force-fit "high self-esteem" onto someone who hasn't earned it, any more than we can force-fit claims of "high quality" onto poor craftsmanship.

A common case of poor craftsmanship is the self-conscious declaration of deferred achievement, for example, by announcing great plans to do something admirable. Author Jeff Haden provides an apt example: a friend in a restaurant says he plans to hike the entire 2,200-mile Appalachian Trail. He talks about equipment he's buying, maps he's collecting, how it will take six months of hard work. Surrounded by amazed admiration, the friend is already basking in the glow of the accomplishment. In fact, he is in an air-conditioned restaurant and will later drive home to his comfy bed.[380]

But will he follow through? Statistically, it is unlikely. Generally, people who announce great plans are much less likely to translate it to action.[381] He will not walk the talk, literally. He will never set foot on the Appalachian Trail. He doesn't have to because he already received the brain-chemical boost from being admired: "this gives the individual a premature sense of possessing the aspired-to identity."[382] He is viewed as a great hiker, or at least he feels he is viewed as such.

Conversely, people who quietly plan are the doers who are much more likely to achieve. They don't announce it beforehand, and they rarely draw attention to it even after they've done it. They do not feel the need to be admired, as their self-esteem is not dependent on how they are perceived, but relies instead on the standards they have set for themselves.

Self-aggrandizement becomes a temporary fix to an unproductive life. This character flaw is typical of adults who grew up with a lot of screentime. The artificial high produced by the dopamine hits that screentime delivers to the brain, can lead to this type of character flaw.

We bask in easy, self-elevating highs from unearned brain-chemical boosts.

The subconscious mind is not fooled by the trick of artificial high self-esteem. The "high" is temporary, since denial of reality, to avoid uncomfortable challenges, will have unpleasant consequences. As clinical psychologist Dr. Bradberry explains, "The biggest obstacle to increasing your self-awareness is the tendency to avoid the discomfort that comes from seeing yourself as you really are. Things you do not think about … can sting when they surface. Avoiding this pain creates problems, because it is merely a short-term fix. You'll never be able to manage yourself effectively if you ignore what you need to do to change."[383]

Authentic self-worth comes from hard work, endurance, character, integrity, and responsibility, to achieve one's goals. It is long-term, authentic, solid, and stable. By applying self-discipline, a positive self-image follows naturally. There is no need to claim it or promote it, or to trick ourselves into it. We accomplish self-worth organically by our actions and by observing our track record.

Conversely, self-conscious emphasis on one's self-esteem produces low self-esteem. It is especially damaging to children. For example, artificially promoting high self-esteem by being overly protective of a child's feelings merely undermines healthy development and ownership of actions. "Unfortunately, without this sense of ownership, children are thoroughly unprepared for the adulthood because in the real world our actions do have consequences."[384]

Exaggerated praise and over-protectiveness produce adults whose self-esteem depends on their egos being stroked by others. "People who are always begging for attention are needy. They rely on that attention from other people to form their self-identity."[385] The result is a fragile and false self-esteem that easily crumbles. The hope, however, is to achieve a more genuine self-worth. As Dr. Bradberry states in his article *12 Habits of Genuine People*, "They aren't driven by ego. … They simply do what needs to be done without saying, 'Hey, look at me!'"[386]

When self-esteem is gained by effort and productivity, there would be no need of accolades in order to feel good. But the more time wasted in passive screentime, the less equipped we are to achieve this quiet road to genuine self-worth.

Dr. Bradberry brings it into perspective: "While it's impossible to turn off your reactions to what others think of you, you don't have to compare yourself with others, and you can always take people's opinions with a grain of salt. That way, no matter what other people are thinking or doing, your self-worth comes from within."[387] Work and achievement feed a virtuous cycle of inner strength and value.

Career Shock

When a parent or a mentor says, "you can be whatever you want to be," there's an opportunity to ask questions, and talk about what it takes to achieve something in the real world. Too often on TV shows and in commercials, "you can be whatever you want to be," becomes a promise in a vacuum. The message implants the passive version of the dream into our consciousness: If I imagine it, it will just happen. It becomes a false hope without the two-way conversation needed to put dreams into action in the real world. The truth is that few ever achieve their early dreams, even by working hard. Dreams change many times throughout one's life and career, with evolving job markets, talents, interests, and other changes. As people are so fond of saying: "Life happens." The needed skill is adaptability, which comes from non-screen real-life experiences.

Children don't have any way of knowing that screentime today will undermine their development for tomorrow. Children don't yet know what it will mean to have their career goals undermined later in life. It is up to parents to provide guidance and protect their children against these shortcomings. As we have seen—in the absence of parental guidance during the formative years—two issues arise from the child's excessive screentime: 1. brain function impairment that reduces opportunities later; 2. the gap widens between screen-fed false hopes and real-world jobs. That one-two punch exacerbates career shock, which contributes to depression, anxiety, and hopelessness.

All too often, college is followed by a bewildering state of uncertainty. Screen-life since childhood left us unprepared and ill-equipped to enter the job market. The jarring clash between pre-career dreams and the mundane struggle to find a job makes a generation start their adult lives in a state of hopeless pessimism, instead of hopeful optimism. Young adults over the generations have always made the transition into full-time employment. The difference today is that screentime displaced real-life-time; and fed us with false expectations.

College Recruiter explains it this way: "Despite all of your job hunting efforts, however, you find that getting your dream job is not as easy as you hoped … in order to get into the job market, you need to take a good look at all of the available choices, adjust your preferences to 'fit in,' and give it your best shot. While it may come as a shock to you, many graduates end up never working in their major. In fact, the majority of graduates don't get to use their degrees at all, and most feel that they are forced to follow a different direction."[388]

While creativity can help advance a career, jobs themselves are typically not very creative. "You'll probably have to pursue your passions during your off-time."[389]

False expectations coincide with excessive screentime, as age-appropriate levels of emotional development lag behind. Painful lessons were not learned over the years, realistic expectations were not set. When screentime was the prevailing entertainment from birth to young adulthood, we never grew the ability to see reality clearly, we didn't experience enough of life to develop basic coping skills, we didn't have struggles or learn fortitude.

The opposite follows from an active nonscreen childhood, where the tools of initiative and emotional maturity equip us with a make-the-most-of-it approach. We have the emotional and intellectual wherewithal to start adulthood with realistic expectations, hopeful optimism, wider opportunities, and strength to cope with obstacles.

Clinical psychologist and author Dr. Travis Bradberry reinforces this idea, "Emotionally intelligent people believe that the world is a meritocracy and that the only things that they deserve are those that they earn. People who lack EQ often feel entitled. They think that the world owes them something. Again, it's about locus of control. Emotionally intelligent people know that they alone are responsible for their successes or failures."[390]

We should expect to work hard, and feel grateful for any chance to get a foot in the door to prove ourselves. We can be better-prepared for hardship and delayed gratification, such as putting in extra hours at night and on weekends for no extra pay or recognition. This personal investment emphasizes personal responsibility for career growth.

We begin with the recognition that promotions may or may not come later after years of proof. As *The Motivation Myth* author Jeff Haden reminds us, "While you ultimately may want to prove that you have the potential to hold higher-level positions, your immediate goal is to be the best in your company at what you *currently* do."[391] To excel in the absence of recognition is a sign of true character.

Achievers will spend their own time and money to gain professional knowledge, to earn the advantage. In his book *Thank You for Being Late,* Thomas Friedman reinforces the point of self-driven improvement, particularly for job markets in technology: "You need to work harder, regularly reinvent yourself, obtain at least some form of postsecondary education, make sure that you're engaged in lifelong learning, and play by the new rules while also reinventing some of them. Then you can be in the middle class."[392] The punchline for Friedman is that this ultra-motivated self-driven pursuit is just enough for an average job and an average middleclass career. Do your very best and don't expect greatness. Friedman talked to hundreds of technology leaders and employers for his book, and characterizes the global competition for jobs in the technology sector as a "brutal meritocracy."[393]

At a minimum, to achieve average middle-class employment in technology, the skills needed include strong "writing, reading, coding, math, creativity, critical thinking, communication, collaboration, grit, self-motivation, lifelong learning habits, entrepreneurship, improvisation."[394] When an opening comes along for a promotion, only those showing initiative in the form of advanced degrees, professional certifications, and relevant product certifications, are considered.[395] If you haven't met these "brutal meritocracy" minimum standards, you won't get the job, or the promotion, and you may never know why you were passed over.

In this new intense world of exponential learning accelerations, for example, "'When I walk into a subway and see someone playing Candy Crush on their phone, [I think] there's a wasted five minutes when they could be bettering themselves"[396] (219). There is no time to waste even a minute. Patience and sacrifice are steady watchwords. Rewards may come in unexpected ways, after years of high performance.

Motivation author Haden continues this theme of self-development: "They prepare. They train. They constantly experiment and adapt and refine, refine, refine. Highly accomplished people gain superior skills not by bursting through the envelope but by approaching and then slowly and incrementally expanding the boundaries of that envelope."[397]

The ability to thrive amid the tough, entry-level reality generates a career path that grows steadily after continual accomplishment. Again from Jeff Haden, "Don't be fooled by the work-life-balance fluff. Tremendous effort and dedication are required.... The only way is the hard way."[398]

The hard way means practice. As Friedman notes, "Practice advances all students without respect to high school GPA, gender, race and ethnicity, or parental education."[399] How do you keep up the practice? Friedman's answer is Concentration: "Students need to learn the discipline of sustained concentration more than ever and to immerse themselves in practice—without headphones on. No athlete, no scientist, no musician ever got better without focused practice, and there is no program you can download for that. It has to come from within."[400]

From the trenches to the top, hard work builds gratifying self-worth and lifelong fulfillment. When you begin to love the struggle as much as the achievement, you've crossed the line into a successful mentality. It's a mentality that leads to meaningful progress.

Builders versus Drainers

There are lots of kinds of people in the world. This section looks at two types: builders and drainers. We have all encountered both kinds of people in our lives. Try this exercise: On a pad of paper, make two columns. Title one column "Builders" and title the other column "Drainers." Write the names of people you know in the column that best characterizes each person. Some may be in a gray area with a mix of personality traits. By and large, however, people fit in one or the other, and we intuitively know the difference.

A builder is someone with initiative who contributes productivity and positive energy to others. They thrive on difficulty, love a challenge, and work constructively all the time. They do this by nature, a builder's nature.

A drainer is a procrastinator who is unproductive and drains energy from others. A key characteristic of a drainer is procrastination, because procrastination impacts others as much as the procrastinator. Drainers delay the unpleasant at both work and home, avoiding the hardest tasks, and others typically have to take up the slack. For the drainer, anxiety grows, stress increases, and consequences mount, as the real-world metes out punishments and penalties, directly and indirectly, for the irresponsible behavior. Everyone around a drainer is negatively impacted, psychologically as well as practically.

Drainers vent at other people as a result of their own failures, which exacerbates a reputation for unreliability and incompetence. Clinical psychologist Dr. Bradberry puts it into perspective: "Every moment spent dreading the task subtracts time and energy from actually getting it done. People that learn to habitually make the tough calls stand out like flamingos in a flock of seagulls."[401]

Builders keep a positive attitude even as they test their limits. They have a positive impact on everyone around them. Their initiative makes them shine.

Builders tackle the worst thing on their plate first every day. As Dr. Bradberry reminds us, they start each day by "eating the frog" (eating the worst item on your plate first, as a metaphor for getting the worst jobs overwith).[402] Builders also display fortitude, and continue the work as long as it takes. They don't wear down or give up.

Author Jeff Haden aptly describes builders this way: "We all have a little voice inside that says, 'I've done enough' or 'I'm exhausted. I just can't do more.' But that little voice lies. We can always do more. Stopping is a choice." A builder's typical day is what Haden calls an Extreme Productivity Day (EPD).[403]

Being a builder does not come easily for everyone. But with practice, the habits of initiative, diligence, and productivity soon become second nature. Drainers spend more time with easy entertainment, such as fun apps, social media, video games, TV, and other passive screentime. Haden's EPD builders do not need to be entertained. Their joy is in the pursuit of goals. Watching the screen does not tempt them. The joys of accomplishment are too great.

Making mistakes is another differentiator between builders and drainers. Everyone makes mistakes. Builders are deeply troubled by mistakes, but they don't waste energy and other people's time making excuses or complaining about it. They admit it and learn from it. Another Bradberry perspective: "You have to make mistakes, look like an idiot, and try again—without even flinching."[404]

If it doesn't go well at first, builders solve problems, recover from failures, and improve. According to Dr. Bradberry, "When hard times hit, people with mental strength suffer just as much as everyone else. The difference is that they understand that life's challenging moments offer valuable lessons. In the end, it's these tough lessons that build the strength you need to succeed."[405]

Bradberry delivers the appropriate ultimatum, "You always have two choices when things begin to get tough: you can either overcome an obstacle and grow in the process or let it beat you. Humans are creatures of habit. If you quit when things get tough, it gets that much easier to quit the next time. On the other hand, if you force yourself to push through a challenge, the strength begins to grow in you."[406]

Builders focus on trying harder, determined to move forward with a purpose. They see the mountaintop and do not hesitate to take the first step. A recent study at the College of William and Mary found that the most successful entrepreneurs can't even imagine failure (even if they've had past "failures"), and pay no attention to naysayers.[407] Both success and failure teach lessons. Builders learn from both and keep growing.

Drainers do very little, but expect very much, and complain that the world isn't fair. Bradberry pulls no punches: "Negative people are bad news because they wallow in their problems and fail to focus on solutions. They want people to join their pity party so that they can feel better about themselves."[408] They stand around idly, amid a plethora of possibilities, distracted by easy screen messages telling them they can be anything they can dream of, but ultimately dream their life away and never building anything. Their world becomes a land of broken dreams and failed potential. Drainers will never live life to the fullest, never put the most into it, and never get the most out of it. Their focus remains in the failure itself, and who else to blame for it. They are unable to get past their lost dream, or to adjust their dreams to fit the reality, to get up, dust themselves off, and pursue new goals through hard work and renewed determination. In short, complaining gets in the way of conquering.

As noted in the earlier section, "Attraction to the Difficult," people throughout history thrived on tough, painful mental and physical challenges. Healthy developed minds feel a strong attraction to difficult hurdles, pursuing them by nature with determination and enthusiasm. These were builders, and they still are today. The difference is, they comprised the vast majority in prescreen eras, but have receded to a small minority in the Screen Age. Drainers have become the majority. This fact has been made painfully clear by the abundant evidence, scientific studies, academic research, and expert commentary already cited in this book.

Delayed Gratification and Redefined Gratification

Builders understand delayed gratification. As *Motivation* author Jeff Haden notes, "Successful people are great at delaying gratification. Successful people are great at withstanding temptation."[409]

A Stanford experiment of a child and a marshmallow tells us all we need to know about the importance of delayed gratification: "There was a famous Stanford experiment in which an administrator left a child in a room with a marshmallow for 15 minutes. Before leaving, the experimenter told the child that she was welcome to eat it, but if she waited until he returned without eating it, she would get a second marshmallow. The children that were able to wait until the experimenter returned experienced better outcomes in life, including higher SAT scores, greater career success, and even lower body mass indexes."[410]

Anyone can have dreams. Anyone can envision greatness. However, success follows only by taking the first tiny steps, knowing that years and miles of solitary discipline must come first. Haden is great at condensing the point to its essential ingredient: "We all *say* we want to achieve things, but we don't really want to achieve them unless we are willing to take the necessary steps to achieve what we say we want."[411] Haden gives us a piece of advice that should appeal to anyone's most basic common sense: "Delayed gratification is always better gratification."[412]

Taking the idea of delayed gratification a step further, we might re-evaluate what is gratification. What about the struggle itself? Successful people welcome a long, hard challenge. They are gratified by it. Yes, they are patient for a future payoff. But they also experience gratification and reward in the immediate arduous tasks. They thrive on difficulty. To succeed, one must learn to thrive on the process, not just the end result. The successful person does this every step of the way, during the years and miles of solitary discipline. It is gratification redefined, and it smells like success, reward, and happiness. The goal remains in the back of the mind, but satisfaction continuously fills the mind and heart from the struggle itself.

It may be a cliché, but is a fact of life: Happiness is a mode of travel, not a prize at the end of the journey. Many have heard this cliché, but few practice it, even fewer live it every minute of every day.

When faced with a journey of 10,000 steps, we can procrastinate that first step forever, because 10,000 seems too hard and too far. Or, we can tackle the first step, even while shaking in our boots, because we crave the challenge and desire accomplishment. The former way sees happiness as too far away, at the 10,000th step. The latter way generates happiness in the first step and in every step thereafter.

Perfectionists

We've heard the adage "practice makes perfect." Societies have long touted the adage as one of the great virtues. In fact, the pursuit of perfection is responsible for much of history's greatest inventiveness, creativity, and productivity. While literal perfection must always remain a little beyond our reach, its pursuit is important for the progress of humanity and the development of individuals.

Having said that, we might be a healthy, productive builder leading an active life, without necessarily being a perfectionist. Perfectionists are builders who take building to the highest level; some would say, a level too extreme. They are the exceptional ones with extraordinary determination. Every excellence achieved is merely another starting point upon which to improve.

Perfectionists have a heightened sense of order and exactitude from the complete big picture to the tiniest detail. Their antennae register the risk of error with extreme sensitivity. When they make a mistake, they don't flinch or break stride. They take the immediate next step to recover from it. They see it as a great opportunity for improvement.

Fretting over a mistake is impossible for perfectionists. They develop ways to move forward rapidly without a blink, striving to get it right and keep doing better. They plow ahead despite opposition, failure, and danger. They have extraordinary emotional intelligence in areas of achievement. "Emotionally intelligent people persevere. They don't give up in the face of failure, and they don't give up because they're tired or uncomfortable. They're focused on their goals, not on momentary feelings, and that keeps them going even when things are hard."[413] This could be said of any builder or active-minded person, but perfectionists take it to the highest level.

Perfectionists take seriously the importance of poise and fortitude under duress. Their sacred duty is to make difficult work become easier through practice and development of expertise—then to move on to something more difficult. To others, their performance seems smooth and easy. That's the intent. The "make-it-look-easy" theme is common when talking about people who excelled in their careers and soared in their disciplines. To cite a couple of examples:

- Dick Van Dyke said of Stan Laurel (of the *Laurel and Hardy* comedy duo), about the artistic effort in his work: "Stan took care to hide it, to conceal the hours of hard creative work that went into his movies. He didn't want you to see that—he just wanted you to laugh, and you did!"[414]

- From a *biography.com* sketch of Fred Astaire: "Fred Astaire revolutionized the movie musical with his elegant and seemingly effortless dance style. He may have made dancing look easy, but he was a well-known perfectionist, and his work was the product of endless hours of practice."[415]

These examples express the idea of perfectionism and poise under pressure. Even if we do not rise to such heights ourselves, they can serve as inspiration for us to do better.

Perfectionist versus Frustrationist

In contrast, some may call themselves perfectionist by wanting everything to be perfect now, without real effort. They complain about imperfections, being frustrated by mistakes, but they don't work to correct them.

You may have heard people say, "I get frustrated a lot because I'm such a perfectionist." The statement itself contains a contradiction. True perfectionists don't spend time getting frustrated, and certainly don't mention it to others.

A frustrationist wishes mishaps for others so that their own imperfect performance may seem better by comparison. This is what we call "Frustrationism," expecting to achieve perfection without effort, with unhealthy responses to disappointments and downfalls. Frustrationism follows from a screen-centered life of lost developmental time, missed opportunities, and stagnation.

This passive-comparison tendency is linked to the artificial-self-esteem tendency discussed earlier—to artificially elevate one's own self-esteem without the painful work of earning it. Comparison also exposes the zero-sum fallacy: Self-image goes up when others go down. The trick of comparison-with-others is a phony substitute for the healthier habit of comparing self with self, before and after. True perfectionists do not compare self to others, but compare self "as is" to self "as could be." The perfect tendency is to push the boundary of our own higher potential. It produces healthy motivation, and fuels genuine confidence, through healthy and intelligent advancement.

Dr. Travis Bradberry frames the idea nicely: "Emotionally intelligent people understand that the happiness and success of others doesn't take away from their own, so jealousy and envy aren't an issue for them. They see success as being in unlimited supply, so they can celebrate others' successes."[416] True perfectionists value genuineness above all—they "want you to do well more than anything else because they're … confident enough to never worry that your success might make them look bad. In fact, they believe that your success is their success."[417] They enjoy surrounding themselves with happy successful people, to learn from and share ideas with.

Perfectionists share a deep desire for all things in life to excel and draw closer to perfection—in oneself, in others, and in all things.

As Bradberry puts it, "Many people approach life as a zero-sum game. They think that somebody has to win and somebody else has to lose. Considerate people, on the other hand, try to find a way for everybody to win. That's not always possible, but it's their goal."[418]

Real perfectionists want a world filled with the highest caliber people, achieving great successes, raising the bar every day. Every step in that direction is celebrated quietly, while striving to do better.

The van Gogh Example

The perfectionist label is sometimes applied to geniuses who suffered mental illnesses, struggled with alcohol or drugs, or endured various kinds of health difficulties and disasters. What about the Vincent van Goghs of this world, who are widely viewed as perfectionists in their specialty, but whose lives are a disaster in other areas?

History shows us many geniuses who focused in their expertise, such as painting, but their lives have not been happy by any standard. Vincent van Gogh is an extreme case in point. But we can still learn something from it.

Vincent van Gogh achieved perfection in his art through his many hours of hard work. He was in and out of an asylum, and ultimately committed suicide. But he created over 2,000 paintings and drawings from 1880 to 1890. About 900 of those were fully realized paintings. That's at least seven paintings and nine drawings per month, every month, for ten years.[419] At today's values, van Gogh would be making about $10 million per month.

He didn't get a penny. He painted because he was a perfect artist—in the sense that he *couldn't not paint*. He was derided and mocked. He suffered in a solitary storm of passion to the death. As a fictional artist of that same era says in a brief dialogue in the *Tragic Muse*, "I see before me an eternity of grinding," "All alone, by yourself, in this dull little hole?"[420] But what comes out from that dull little hole seems miraculous.

Imagine if van Gogh's developmental years were replaced with screentime like most people today. From the studies and research we have seen, van Gogh's mind, energy level, skill, and creativity, would all have deteriorated before he had a chance to put his first paint brush to his first canvas. We would not recognize his name today. To our benefit, he lived before the Screen Age.

We think highly of those who have contributed to society through the perfectionist's extreme drive to create, even to the point of madness. But do we want to emulate van Gogh's lifestyle? No! Can we learn something from van Gogh's dedication? Yes!

While many say there is a thin line between genius and madness, we needn't descend into madness in order to achieve excellence. Inspiration is there for us to seize and advance in our own way. With the natural discipline and drive that served us well throughout history, we can rise above today's empty screen-induced inactivity, achieve great things, and contribute to others.

CHAPTER 8: THOUGHTFULNESS AND WISDOM

Thoughtful can mean both "kind" and "thinking." Historically the two were closely related. It was the sign of a thinking person to act with appreciation and gratitude, to practice good manners, to show sincere courtesy towards others. It is obvious that courtesy has diminished markedly in our society. As clinical psychologist and *Emotional Intelligence 2.0* author Travis Bradberry notes, "With the decline of good manners, there are fewer expressions of appreciation."[421]

Courtesy evolved as a vital aspect of wisdom, because thinking people understood the far-reaching personal and social benefits of a thoughtful and courteous population. Thinking through to the logical conclusion, if courtesy and thoughtfulness were the rule rather than the exception throughout the population, we would enjoy being around other people more. There would be less stress and more camaraderie. Enjoying other people's company has suffered from our screen-based lives.

In fact, enjoying other people's company is undermined by the mere presence of a smartphone, according to a study involving over 300 community members and students to share a meal at a restaurant with friends or family. "Participants were randomly assigned to keep their phones on the table or to put their phones away during the meal. When phones were present (vs. absent)" Participants with the phone present did not enjoy the time; participants without phones present enjoyed the time. We found consistent results…during in-person interactions, participants reported lower enjoyment with phones than without phones.[422]

Enjoying other people's company always puts the emphasis on the other person. Focusing on people we care about increases enjoyment. Phones undermine our ability to focus on the person we're with.

A completely different study found results that may be related. Reported in The University of Chicago *Journal of the Association for Consumer Research*, the study found that having a smartphone in the room lowers brain capacity, even when the phone is off and out of sight.[423] Consciously or subconsciously, the brain uses energy to resist picking up the smartphone. This highlights the power of screen addiction that most of us are not even aware of. Even when the smartphone is off, and out of sight (but still in the room), our brains are fighting the impulse to use it. The study showed that the solution is simple, leave the phone in another room. This tactic was consistently successful. Brain capacity returned to normal simply by not having a phone in the room. Both enjoying people's company, with less stress, and functioning at full brain capacity, depend on not having a phone nearby. Separating from our smartphones is a simple way to improve quality of life in many ways.

Thinking clearly and enjoying people's company lead to more pleasant interactions—a more polite and friendlier social life. More polite social encounters happen naturally when we place the emphasis on each other.

Dr. Bradberry explains the emphasis on others, "A lot of people have come to believe that not only are manners unnecessary, they're undesirable because they're fake. These people think that being polite means you're acting in a way that doesn't reflect how you actually feel, but they've got it backwards. 'Minding your manners' is all about focusing on how the other person feels, not on how *you* feel. It's consciously acting in a way that puts other people at ease and makes them feel comfortable."[424]

Discourtesy is a self-indulgent behavior symptomatic of a screen-centric lifestyle. The symptoms arise from both the emotional damage, disengagement, and isolation caused by screentime.

Courtesy, on the other hand, would improve everyday life for everyone. We would look forward to simple activities like buying groceries, driving to work, stopping at the bank, shopping for clothes; because those activities involve seeing people, who would be a pleasure to interact with as a rule. This healthier and happier life is a fact of brain chemistry and the positive feelings we receive from positive encounters.

We would be more willing to take our free-time activities out around other people: going to parks, walking in the neighborhood, sitting outside saying hello to neighbors passing by, walking downtown, visiting the library, going to outdoor markets, participating in community events. Courtesy and camaraderie shine a brighter light on every social activity.

It almost seems like a dream of yesteryear to many, the ideas of enjoying meeting people, enjoying talking with "strangers" out in public every day. It was normal and common for thousands of years.

We can still learn from the 1340s cleric who said of Paradise, "There one learns to live well in wisdom and courtesy, for no villainy enters there."[425] It reminds us of the vital connection between wisdom and courtesy, and how much screenlife has made us forget. Who would not rather see a restoration of that connection? Why wouldn't we want to restore a sense of community, neighborliness, enjoying each other's company?

Clinical psychologist Travis Bradberry summarizes the practical benefits: "Being considerate is good for your mental and physical health, your career, and everyone around you."[426] Screen addiction remains our primary obstacle.

Listen to Communicate

There are a lot of ways to communicate. The best way is to listen. We can tell by the response when someone was not really listening. They reply with something that is off topic or incongruent. They ask a question that was just explained. Or they just ask us to repeat the whole thing because they "didn't catch it." Often the reply switches the focus over to the other person—the listener is sending a message: "Whatever you said, it's more interesting if we make it about me."

Most of us have experienced this, and we probably felt disappointment from a distracted or self-centered listener. In these cases, there is a subtle but real message of disrespect towards the speaker. As clinical psychologist Dr. Bradberry notes, "The biggest mistake people make when it comes to listening is they're so focused on what they're going to say next or how what the other person is saying is going to affect them that they fail to hear what's being said."[427]

It's worse when we let an electronic device blatantly divide our attention. As Dr. Bradberry notes, "Nothing will turn someone off to you like a mid-conversation text message or even a quick glance at your phone. When you commit to a conversation, focus all of your energy on the conversation."[428]

Conversely, it is rewarding and refreshing when a listener responds with an intelligent question that demonstrates attentiveness. Dr. Bradberry encourages this kind of attentiveness: "When someone is talking to you, stop everything else and listen fully until the other person is finished speaking … pick up on the cues the other person sends, and really hear what he or she is saying."[429]

Aside from the courtesy element, there is a self-improvement element. People who learn the most, listen the best. As Hemingway noted, "I like to listen. I have learned a great deal from listening carefully. Most people never listen."[430]

Happy Minded

An active mind makes us happier. The active-happy correlation is a strong one that has been well known for hundreds of years. As British Prime Minister Benjamin Disraeli said in 1884, "Action may not always bring happiness; but there is no happiness without action."[431] That goes for both mental and physical action. Aristotle puts it simply: "Happiness is a state of activity."[432] By implication, watching a screen is a state of unhappiness. In fact, as a study reported in the *Journal of Economic Psychology* makes clear, TV viewing decreases life satisfaction, and increases anxiety.[433]

In a study tracking the well-being of teenagers, and how they spend their time, since 1991, "every activity that didn't involve a screen was linked to more happiness, and every activity that involved a screen was linked to less happiness." But what about the reverse cause-and-effect? Perhaps people are unhappy first, and unhappy people gravitate to the screen. The studies ruled out the reverse cause-and-effect. As San Diego State University Professor of Psychology Jean Twenge reports, "Several longitudinal studies show that screen time leads to unhappiness but unhappiness doesn't lead to more screen time." We know that screentime is the cause—it undermines well-being and happiness.[434]

Like so many other studies we've already cited in this book, this study also found that teens' happiness suddenly plummeted after 2012 as smartphone use grew rapidly. Several studies mirror the pattern: The spike in smartphone use was followed by a spike in teenagers' major depression, self-harm, and suicide.[435]

Conversely, discontinuing screentime immediately improved well-being and happiness. Advice based on the research is clear—to be happy, put away the screen device, and do almost anything else.[436]

Part of the problem with screentime versus happiness may be that we underestimate how much we would enjoy spending time with our own thoughts, away from screen distractions. New research published in the *Journal of Experimental Psychology: General*, shows people do not accurately predict their own enjoyment level from unplugging and just thinking, versus the enjoyment of using a screen device. In varied environments, for varied amounts of time, in all cases, people enjoyed spending time with their thoughts significantly more than the people themselves had predicted. Study co-author Kou Murayama, PhD, of the University of Tübingen in Germany, notes that the results are especially important in our era of constant access to distractions. He said, "It's now extremely easy to 'kill time.' On the bus on your way to work, you can check your phone rather than immerse yourself in your internal free-floating thinking, because you predict thinking will be boring."[437]

Making that first break from screentime, to do something else, anything else, seems to be the biggest obstacle. But if one fact is obvious, it is that happiness depends on removing or significantly reducing screentime.

So what is this illusive condition called "happiness" that we lost over the past forty years? What are happy people like?

If you've ever met one, you will find that people who seem happiest, seem happiest no matter what happens to them, good or bad. As noted by experts in this area, Drs. Jean Greaves and Travis Bradberry, "Even when you can't do or say anything to change a difficult situation, you always have a say in your perspective of what's happening, which ultimately influences your feeling about it."[438]

For an unhappy-passive mind, a little trouble generates a lot of bitterness. A little success breeds a lot of gloating and exaggerated celebration. An unhappy-passive mind handles both success and failure poorly, and learns from neither.

For a happy-active mind, a lot of trouble does not cause bitterness, but increases fortitude. Success breeds no gloating, but more humility and perseverance to succeed more. Active-happy-minded people handle both success and failure with grace and learn from both.

Active minds are happier because they process incoming data more effectively. That in itself leads to healthier brain-chemical reactions. It is so simple, it becomes a matter of statistical predictability. From a 2012 *Ted Talk*: "if I know everything about your external world, I can only predict 10 percent of your long-term happiness. 90 percent of your long-term happiness is predicted not by the external world, but by the way your brain processes the world."[439] Screentime damages your ability to process the world.

Active minds naturally generate happiness as a mode of travel through life. A Passive mind doesn't process life in a healthy way, so it inevitably views happiness as an object to obtain. Grasping for happiness like an object is the most famous road to disappointment and bitterness.

Wise people have known that happiness is a mode of travel, and a way of processing, since the beginning of time. Founding Mother Martha Washington, for example, said "for I have also learnt, from experience, that the greater part of our happiness or misery depends upon our dispositions, and not upon our circumstances. We carry the seeds of the one or the other about with us, in our minds, wheresoever we go."[440]

Viktor Frankl was a concentration-camp survivor during the WWII Holocaust. There is perhaps no better authority on choosing our response to circumstances. Frankl points out that "people often react without thinking. We frequently don't choose our behaviors so much as just act them out. But [Frankl] observes that we don't need to accept such reflexive reactions. Instead, we can learn to notice that there is a 'space' before we react. He suggests that we can grow and change and be different if we can learn to recognize, increase, and make use of this 'space.' With such awareness, we can find freedom from the dictates of both external and internal pressures. And with that, we can find inner happiness."[441]

When we remove screentime, and replace it with almost any kind of mental or physical activity, we immediately start healing our brain's processing apparatus. We improve our happiness. We switch from frustrated grasping, to successful living. Franklin Delano Roosevelt provides our last comment on this topic: "Happiness lies in the joy of achievement and the thrill of creative effort."[442]

Only Human versus Fully Human

When a passive mind meets a high standard, he says "Nobody can measure up to that, it's not human!"

When a drainer meets a builder, he says "We're not all Superman! He's not human!"

People justify a low-caliber lifestyle by saying "I'm only human!"

Being "only human" means being incompetent, unreliable, weak, and careless. It means that humans who try harder are not human because they try harder. This is the prevailing attitude in a screen culture of passive entertainment, and suppressed development.

The truth is, however, achievement makes the more disciplined person live more fully. As *The Motivation Myth* author Jeff Haden bluntly puts it: "If I haven't convinced you that being a serial achiever is the best way to live a full, satisfying, and successful professional life and personal life … well, there's no hope for you."[443] By being active, strong, creative, reliable, and competent, the disciplined person is more fully human.

Easy versus Difficult

Here are two important takeaways:

1. Easy is empty
2. Difficult is rewarding

Mistakes can produce two kinds of responses:

1. Negative response: "I'm only human (I accept the lesser me)"
2. Positive response: "I'm fully human (I demand a better me)"

A Different Kind of Smart

Intelligence as measured by typical assessments does not guarantee success or even make it more likely. A passion to persevere and improve has a stronger correlation to success. It is an active-minded characteristic.

Standard measures of intelligence tell nothing of the quality of person—or quality of life—compared to the measure of an active mind versus passive mind. That is why Emotional Intelligence (EQ) is so much more important than IQ. There are two import practical differences between IQ and EQ, as described by *Emotional Intelligence 2.0* author and clinical psychologist Travis Bradberry:

1. IQ is an innate characteristic that cannot be increased; but EQ is a skillset and a discipline that we can choose to develop.[444]
2. EQ is a much stronger indicator of success than IQ.[445]

This is an amazing gift! The tools for success and happiness are within our control, if we want it badly enough. The first building block is to establish daily active-minded habits. The choice is ours to embrace.

- High EQ is the hallmark of the active mind.
- Low EQ is the hallmark of the passive mind.

Psychologist Angela Duckworth spent years studying what makes high achievers so successful. Her results provide us with useful advice that we can put into action:

> "It wasn't SAT scores. It wasn't IQ scores. It wasn't even a degree from a top-ranking business school that turned out to be the best predictor of success. 'It was this combination of passion and perseverance that made high achievers special,' Duckworth said. 'In a word, they had grit.' Being gritty, according to Duckworth, is the ability to persevere. It's about being unusually resilient and hardworking, so much so that you're willing to continue on in the face of difficulties, obstacles and even failures. It's about being constantly driven to improve. In addition to perseverance, being gritty is also about being passionate about something. For the highly successful, Duckworth found that the journey was just as important as the end result. 'Even if some of the things they had to do were boring, or frustrating, or even painful, they wouldn't dream of giving up. Their passion was enduring.'"[446]

The key to this story is having the passion to push through boring and painful work. Passion is not about joy. It's about pain.

In another study, "only 25 percent of job successes are predicted by IQ, 75 percent of job successes are predicted by your optimism levels, your social support and your ability to see stress as a challenge instead of as a threat."[447]

In another example, Jon Morrow, an all-around thinker who is well-known in the world of blogging, provides a good example of active-minded smarts trumping conventional IQ measurements, in two of his 2014 posts ("How to Be Smart in a World of Dumb Bloggers" and "On Gluttony, Selfishness, and Unleashing the Power Within").[448] His self-avowed average IQ by standard measures had no negative impact on his rise to the top in his business. His passion, dedication, and purposeful development resulted in success. "Each and every popular blogger I know spends at least three or four hours a day consuming new information. It's not just an idiosyncrasy. It's *required.*"[449]

All of these studies and examples lead to one conclusion: The best kind of "smart" is active-minded Emotional Intelligence (EQ), away from passive screen stagnation.

Intuition

Intuition is a science-based subtle brain activity that incorporates conscious and unconscious messages. "Even if you're not consciously thinking about it, your unconscious brain starts working on a problem or decision right away." High emotional intelligence (EQ) plays an important rôle.[450]

An integral part of happy and healthy living is the ability to make sound decisions. The more thought we bring to decisions, the better. We are complicated organisms, so it is not easy to take care of our competing needs when we have to make an important decision. How do we balance emotional needs, rational principles, and practical considerations? Intuition is an important part of that equation.

For example, following only emotions usually leads to unhappiness, when we ignore practical considerations, which bites us later. Purely practical considerations can also lead to unhappiness, if we neglected an emotional requirement about who we are. Emotional and practical considerations are important, but not enough alone. The critical ingredient that puts both into perspective is intuition.

Intuition is a deep awareness of both external surroundings and internal self. It is a more complex faculty that includes both emotional and rational awareness, with both unconscious and conscious inputs. As author Jeff Haden highlights in a recent article, Nobel Prize-winning economist Daniel Kahneman explains that "Intuition is thinking that you know without knowing why you do. Like quickly scanning three resumés and, without thinking too hard, picking the best candidate. Like quickly scanning long checkout lines at the supermarket and deciding, without thinking too hard, which is likely to be the quickest."[451]

Other examples might include a prediction, an idea that a friend has a problem, a sense of knowing that we'll like someone when we first meet them, a decision we have to make at work when there is not adequate information, knowing which way to turn at an unfamiliar location, a decision about a vacation, a sense that we will like this job but not that one, an insight in a conversation, studying a new subject, a decision on buying a house. Intuition is, in a word, awareness. Awareness is a deeper sense of attention that never develops when we are distracted by screentime.

As clinical psychologist Dr. Bradberry explains, "Awareness of yourself is not just knowing that you are a morning person instead of a night owl. It's deeper than that. Getting to know yourself inside and out is a continuous journey of peeling back the layers of the onion and becoming more and more comfortable with what is in the middle—the true essence of you."[452]

Often we may see the right path or the right decision for us, but we give in to outside pressures and lose confidence. Perhaps the most important criterion for our own fulfillment, after making the best possible decision, is to trust it. Don't give in to other people's doubts and criticisms. Awareness means mental strength that values our own intuitive thinking: "People with mental strength believe in themselves no matter what, and they stay the course."[453] And As Nobel Prize-winning economist Daniel Kahneman notes, "Granted, sometimes you'll be wrong. But, depending on your level of emotional intelligence, not as often as you think."[454] Turn to intuition more than mere emotions and mere practicality for guidance, then trust it.

The more important the decision is, the more important intuition becomes. Many life-changing, life-long choices are made while we're very young; career and marriage, for example. Childhood screentime robs us of intuitive decision-making skills that are critical in young adulthood. Failing that development, a lot of bad choices and regrets will follow.

Intuition doesn't necessarily equate to that unsubstantiated inclination that we often call a "gut feeling." Following a momentary gut feeling and just going with it, without reflection, can be just as misguided as ignoring intuition altogether. Instead of yielding intuitive insights, gut feelings are often no more than rapid surface decisions when we're too lazy to do our homework.

Again as author Jeff Haden highlights in a recent article, Neuroscientist Friederike Fabritius and Psychology PhD Hans W. Hagemann point out that "Intuitive decisions are often the product of years of experience and thousands of hours of practice. They represent the most efficient use of your accumulated experience."[455] Intuition depends on those thousands of hours experiencing life, time spent reflecting, and deeper levels of concentration. These cognitive processes are undermined by screentime—intuition is undermined by screentime.

The Aged and the Handmade

One way to deepen life experience is to find some good old-fashioned hidden beauty in the asymmetrical, individual, authentic, quirky, natural characteristics of people and objects. Screentime removes this kind of experience from our lives. This beauty emphasizes authenticity, behaving in natural ways, uniqueness of appearances, appreciating diversity of talents, valuing different styles of expression—the expressions of people and nature in person.

These older concepts of beauty have been largely forgotten in the current era of virtual reality, sparkling HD screens, laser-precision industry, and special effects. Older concepts of beauty require long attention spans, focused concentration, reflective patience, close observation in the natural, three-dimensional world. They do not occur on a screen.

This kind of beauty celebrates the imperfect, unique, understated, authentic qualities of people and things. It emphasizes characteristics that are not obvious. It might be the beauty of a crack in a table that brings back fond memories of an event thirty years ago, or an imperfection in someone's face that makes the person more attractive and interesting. Older concepts of beauty are the opposite of CGI and polish.

Another example might be the sparse beauty of a small home with a nonsymmetrical meticulous interior, a few well-made pieces of essential furniture, with very tasteful but very little decoration.

Authentic beauty shines in the marks of aging in a person, or the effects of the elements on weathered objects. It could be the character in an aged person's face and hands, or the glow of a 90-year-old man's understated smile playing off his weathered skin. It might be an old pair of faded, torn, and patched blue-jeans. The emphasis is authenticity, so a brand-new pair of pre-faded, or pre-torn blue-jeans off the shelf would be a mockery of this kind of beauty—phony expression—like fake self-esteem. Faded, worn, torn clothes from the wearer's activity over the years becomes an expression of that individual.

Old-fashioned beauty is also found in hand-crafted products. Factory-output products generally have geometrically even and symmetrical designs. Conversely, handmade tables, chairs, carpets, cabins, clothes, blankets, pottery, and most anything else handmade, show the uneven results inevitable with different individuals. We see expressions of design, manually crafting, carving, coloring, building, weaving, sewing, molding, and assembling. These reflect the uniquely human act of creation.

The least human types of creation are AI-generated simulacra of text, art, and other objects. They can write your novel for you, or your term paper, or populate your next exhibition, but AI is just another piece of "smart" software that makes people dumber.

The Aged and the Handmade types of expression, however, give us a deeper insight into the human spirit—a dimension of meaning to life that is largely absent today.

PART III: Epilogue

CHAPTER 9: THE DYING OF THE LIGHT

In this chapter we explore the long-term losses from prescreen eras, which cannot be recovered in the Screen Age.

TV Drains the Color Out of Life

People used to say the 1960s is when the world became more colorful (applying a metaphor of color-TV technology and colorful 1960s styles). But the truth is, because of color TV, the 1960s is when the color drained from the world, and we the people faded to black-and-white. As TV changed from gray to color, we changed from color to gray.

Today we mourn the loss of strength, the loss of wisdom, the loss of creativity, the loss of discipline, the loss of inventiveness, the loss of initiative, the loss of camaraderie. Screen immersion is like a poison that deadens these areas, while temporarily stimulating pleasant sensations. Every video, every TV show, every minute of screentime, deadens our hearts and minds a little bit more.

Anxiety about Death: Where Did the Time Go?

In the past, by age 75, people had gathered a wealth of life experiences and had no anxiety about dying of old age. They lived a full life, and they felt it inside. We experience something very different today. By age 75 we will suffer terrible anxiety at the thought of dying, because to us it seems like we only lived twenty-five years. "How can I be nearing the end so soon?!" In a way, that impression is correct. People of the Screen Age really lived only twenty-five of their seventy-five years. At age 75 we will have the life experience that a 25-year-old had in the past. We spent fifty of those years awash in screen images, experiencing nothing of substance.

In the past, if we asked 300 million people what they did with their lives, we got 300 million different answers. Today when we ask 300 million people what they did with their lives, we get one answer: "I watched everything on a screen."

Where Are the Greats?

Where are all the greats like those we hear about from the past? Where are today's versions of yesterday's Faulkners and Hemmingways and Angelous, the Benjamin Franklins and Mark Twains, the Albert Einsteins and Thomas Edisons, the Vincent van Goghs and Leonardo da Vincis, the George Washingtons and Abraham Lincolns, the Martin Luthers and Martin Luther Kings, the Mozarts and Bachs, the Louis Armstrongs and Benny Goodmans, the Louis Pasteurs and Jane Goodalls, the Andrés Segovias and Isaac Sterns, the Ginger Rogers and Fred Astaires, the Buffalo Bills and Annie Oakleys, the Orson Welles and Alfred Hitchcocks, the Arthur Conan Doyles and Agatha Christies, the Eddie Rickenbackers and Amelia Earharts, the John Muirs and Walt Disneys?

The answer is: They grew up with screens since infancy, immersed in screens through childhood and adolescence, and became lifelong screen-addicted adults, achieving very little. Today's existences will never become an inspiration to others. Mass assimilation of screen passivity has replaced individual achievement.

People who might have become great did not lose the potential by choice. It happened in the natural course of screentime replacing development time. There was no choice. Day by day, hour by hour, screentime systematically drained the strength and hope out of those who might have become great in the past fifty years. Upon a once-adventure-filled road of life, the roadside is now littered with decaying skills, dead potentials, unfinished accomplishments, dreams up in smoke, culture in ashes.

Even early TV stars note that they honed their craft because they had to entertain themselves in their own prescreen days. For example, TV icon Dick Van Dyke recalls, "We got together and harmonized, told jokes, and invented tall tales that kept us amused. We spent hours exercising our imaginations and entertaining ourselves in those days before television and long before the Internet."[456] TV stars with prescreen childhoods look around them, and don't like what they see in their own profession. *Columbo* star Peter Falk, for example, talked about, "my contempt for much of what is seen on TV—my refusal to deal with mindless action, dumb jokes, and exposition posing as real talk."[457]

Literary figure Philip Roth made similar observations. Roth lamented the decline in the general population's literacy that he witnessed during his lifetime. He noted that, among writers, he and John Updike were "the last pre-television generation."[458] As television implacably dumbed down the populous over the years, there followed "the inevitable decline of 'people who read serious books seriously and consistently'" … "Someday soon, said Roth, reading novels would be as 'cultic' an activity as reading 'Latin poetry.'''[459]

As the years go by, we have to keep looking farther and farther back to find anything developed and well done, let alone heroic or brilliant. Everything is reduced to mediocrity. Lions of public service and the arts are relegated to history books and museums, which are almost never consulted or visited. They have an alien feel about them, unfamiliar and difficult to imagine. This endangered species will soon become extinct without much notice.

Where Are the Wise Elders?

Wisdom survives and grows only when all of us work, play, and live together, in person. Given our amount of screentime today, we cannot possibly retain the wisdom of past generations. We certainly are not adding to it. Cumulative passive screentime for the whole population during the past sixty years equates to millions of lost hours of engagement with other people, including our most precious resource: Elders.

The only way to circulate the life learnings of older people is to pass it down to younger people, in person, so they can keep it alive and keep passing it down. Otherwise, nothing is preserved or built upon, and lifetimes of knowledge fade into extinction.

TV in the 1950s and 1960s marked a new phase when screen culture began cutting off that circulation of knowledge. Today's screen devices finish cutting it off, as our immersion in screens fully displaces the human connections required to sustain the circulation of knowledge and wisdom. Ironically, the smartphone is the gravestone that marks the dead body of wisdom that once circulated in the veins of humanity.

As we—the unwise screen-centered people—grow older, we have much less to pass along to the next generation. Each generation diminishes, gradually ending near zero. We cannot miss what we never noticed. Generations of knowledge and wisdom die away un-mourned and unnoticed. As we relinquish our energy to the magnetic attraction of the screens all around us, we lose priceless treasures and gain nothing.

Instead of being mourned, our animated and vibrant ancestry is typically deprecated, denied, resented, laughed at, or dismissed by today's a shallow house of fools who skim the surface of life through screens instead of living. That is our mocking tribute.

Historia Magistra Vitae Est

Using Latin may sound pretentious to some. But it's a good way to reinforce a point. As we lay to rest our dead elders, philosophers, prophets, poets, and all of our ancestors, we lose our deepest springs of life, we lose golden insights by the thousands. We lose diligence. What's worse, we lose the joy of diligence, which comes from the Latin *diligens*, *diligere*, meaning to love and to esteem: the love of learning, and the esteem of learning. We replace a rich, deep expression and experience with shallow screen distraction that keeps us dull and thoughtless.

Screenlife begets mediocrity in our society and in ourselves. We lose our best selves, and settle for less. And so we turn to a Latin quote from one of our dead elders, Cicero, "Historia magistra vitae est"—History is the teacher of life—and we have lost our history.[460]

The Myth of the Myth

Some people claim a book like *Screenformation2.0* is inventing a mythology: The myth of a greater civilization that is now lost, a romantic notion of a stronger, smarter, nobler, more capable species that roamed the earth in the legendary past.

If only it was a myth, we could breathe a sigh of relief. Unfortunately, we now have overwhelming scientific evidence that demonstrates how screentime has damaged our society, as well as our individual brain function, mental health, emotional development, and physical health.

Choosing our status quo screen-based lifestyle means choosing emptiness, frustration, and depression, ultimately towards self-destruction. It is a path of denial, embracing the comforting lie that nothing's wrong.

The false story is the one that says we are advancing and evolving in a positive way. Screens all around us feed us the comforting lie that we progress and advance with each new generation. The lie feels so good, we seem to *have no choice* but to believe it. Because of screen-suggestibility, "feels good" equals "true." The absurd advance-as-a-species story, however, is the most dangerous lie, and the most fraudulent mythology. We remain a captive audience in our prison of lights, the screens that surround us. In it, we are bombarded by the lie in a thousand ways every day until accepting it is as natural as taking our next breath.

Hope for the Future?

While the bleakness hovering over our collective spirt in the form of fifty billion screens is not likely to brighten in the near future, there is still hope for individuals. Anyone can decide to be an exception to the rule by cutting out the screen from our own daily habits. Replacing passive screentime with active participation in life gets easier after a little progress. *The Motivation Myth* author Jeff Haden's insight is not oversimplified, it is just profoundly simple: "A productive body in motion tends to stay in productive motion."[461]

There are tactics to help switch from passive stagnation to active achievement gradually. Focusing on one small change at a time, repeating active habits with tiny increases over time, attaching a new good habit to an existing routine ("stacking habits"), are ways to get started (from *Exist* co-founder Belle Beth Cooper's story).[462] Increasing real-life activities in itself will begin to displace screentime.

Whether we unplug completely, or find healthy ways to limit screentime, our biggest challenge is to *do something*. One of our most important choices in life will be whether to settle for the "as is" self of being "only human," or rise to the "can be" self to become "fully human."

Final Message for Young Parents

Success comes from taking the high road of strength and action. For parents especially, consider the tragedy of wiping out all of your child's real-world and rewarding experiences and life skills, and replacing them with incompetence and shallow screen-memories.

Step back and look at the broader view. You are equipped with the scientific proof of screentime's harmful effects on child development (as this book has shown beyond question). Observing typical American households, you see two 30-year-old parents and their 3-year-old child. Each day of the year, those parents must make a choice:

A. Develop the child's brain function, mental health, emotional development, and physical health; or
B. Damage the child's brain function, mental health, emotional development, and physical health.

Now ask the question about those two 30-year-old parents: Each year, how many of the 365 days do you think the parents should choose A? How many of the 365 days do you think the parents should choose B?

Now step back into your own life. Your child's potential is in your hands: A or B? Every day. The choice is that simple.

CHAPTER 10: POSTSCRIPT

Author's Notes

My first exposure to studies on TV's damage came from reading Dr. Marie Winn's *Plug-In Drug* in 1981. But long before I read *Plug-In Drug*, I had already stopped watching TV. I picked up Winn's book out of curiosity to see if my concerns about TV aligned with the results of studies and research presented in her book.

It did. My impression of TV was confirmed by scientific study more than forty years ago, and by countless more studies since then. Common sense told me that screentime displacing active time must be harmful. I didn't know yet that scientific studies and academic research over subsequent decades would prove screentime to be far more harmful than I had suspected. After a moment's reflection, however, I think we all have the same inkling of the screen's unhealthy effects. Haven't we admitted it every time we've said "I watch too much TV"?[463]

Very often we watch a lot of TV, but say "I really don't watch that much TV." We really believe it, because TV is such a thief of time. We watch TV for five hours, and the next day we swear that we only "watched about an hour of TV last night." The other two or three hours of free time were spent on social media and other screens, mostly consuming video on several devices. Screentime is virtually unconscious time, except we are soaking in countless messages that quietly shape our brain during that evaporated time. Hence the screenformation of our minds and our society.

Author Biography

Robert Rose-Coutré has been studying people and taking notes since earliest memory. Starting at a young age, he spent his life studying people, society, history, and other cultures. Since early childhood he sought out older people to learn as much as possible about older generations. Some of his earliest boyhood memories included talking with World War I veterans or anyone born in the 1800s to learn what life was like in the old days. Those experiences brought home how precious people are, that the old-school ways of fully-human living should not be dismissed and lost so easily. His motivation has been to learn about human feeling, thinking, and behavior; understanding relationships, appreciating people, and leading a meaningful life that benefits others.

Robert Rose-Coutré has a Master of Arts (MA) degree in English Language and Literature/Letters. He studied Latin, philosophy of language, epistemology, logic, analysis, aesthetic theory, German, film theory and criticism, and music theory.

Rose-Coutré has been a member of high-IQ societies such as American Mensa, Intertel, ISI-Society, World Intelligence Network, and the Ronald K. Hoeflin Societies. But high-IQ-society membership is meaningless without the practical application of the those abilities channeled constructively. *Screenformation 2.0* gives us the practical application. The book provides knowledge, guidance, and motivation that readers can apply to everyday life.

Careerwise, Rose-Coutré has been a digital marketer, editor, writer, and publisher. In his personal writing endeavors, he previously published a book of commentary on active living; a scholarly book on works of art, abstract objects, and the philosophy of language; and a novel. His specialty, as demonstrated in *Screenformation 2.0*, is in scientific and technical research, communicating complex information clearly and accessibly for the general reader.

Robert Rose-Coutré is grateful to have had a TV growing up, having watched enough to gain an intimate understanding of its effects, but limited enough not to do too much damage. He is grateful to his parents for those strict limits. Because of that limitation, he had time for activities that leave him with precious memories of old-style living.

With those extra 20,000-to-30,000 hours of childhood and teenage screenfree development, he grew up doing real-life activities: read every volume of his family's encyclopædias along with a lot of fiction and nonfiction books on culture, science, and history; played long hours of football, baseball, basketball, soccer, tennis, and golf; sold newspapers at a shopping center at age 10 and had many more jobs growing up; did yardwork and household chores most every day, always on Saturday;

learned household maintenance like replacing fixtures, roofing, dry walling, painting houses interior and exterior, welding, metalworking, woodworking; rode a bicycle everywhere many miles a day including to and from school and jobs (parents didn't give rides); became a competitive speed skater at the local roller-skating rink at age 10; got in fights with kids in the neighborhood—learned the difference between healthy fighting versus vicious violence; started a daily exercise regimen at age 13 (and kept it up ever since); learned how to play the organ and piano from his grandmother (who had performed in early Vaudeville and in Chicago nightclubs); took guitar lessons; sang in the church choir; won or placed in small local tournaments in chess, billiards, and ping-pong, and played those games regularly; learned sailing, motorboating, rowing, canoeing, and water-skiing; learned martial arts, juggling, knife-throwing, target shooting, archery, and horseback riding; learned endurance swimming and lifesaving and became a lifeguard at age 16; camped on weekends once a month and learned skills in camping, wilderness survival, and many other outdoors skills; worked through the ranks of scouting and earned Eagle Scout; participated in community service projects through scouting and church; listened to his parents' and grandparents' stories of their childhoods; and enjoyed almost all recreation time with friends, doing things in the real world with real people, growing and learning how to be a human, an old-school human, fully human.

Today too many of us miss out on the excitement of life, trading it in for a lifetime of passive screentime. We replaced curiosity, passion, and action; with boredom, stagnation, and shallow entertainment. But we can still choose the active, thoughtful path of curiosity, passion, and accomplishment. Reject the screen and join the living.

Links to Companion Pages

- Website:
  https://www.screenformation.com
- X.com:
  https://x.com/screenformation

PART IV: Works Cited and End Notes

WORKS CITED

Abi-Jaoude, Elia, MSc MD; Karline Treurnicht Naylor, MPH MD; and Antonio Pignatiello, MD. "Smartphones, social media use and youth mental health" US National Library of Medicine, National Institutes of Health, *Canadian Medical Association Journal*, Vol. 192(6): E136–E141. https://www.ncbi.nlm.nih.gov/pmc/articles/PMC7012622/, February 10, 2020. https://www.cmaj.ca/content/192/6/E136, February 10, 2020.

Achor, Shawn. "The Happy Secret to Better Work" *Ted Talk*. https://www.ted.com/talks/shawn_achor_the_happy_secret_to_better_work/transcript?language=en, February 2012.

Alexander, Christopher; Sara Ishikawa; Murray Silverstein. *A Pattern Language*, Oxford University Press, 1977.

Alexander, Patricia and Lauren Singer Trakhman. "The enduring power of print for learning in a digital world" *The Conversation*. https://theconversation.com/the-enduring-power-of-print-for-learning-in-a-digital-world-84352, October 3, 2017.

Alexander, Patricia and Lauren Singer Trakhman. "Reading Across Mediums: Effects of Reading Digital and Print Texts on Comprehension and Calibration" *The Journal of Experimental Education*, Pages 155-172. Published online: https://doi.org/10.1080/00220973.2016.1143794, March 9, 2016.

Alter, Adam. "Tech Bigwigs Know How Addictive Their Products Are. Why Don't the Rest of Us?" *Wired.com*. https://www.wired.com/2017/03/irresistible-the-rise-of-addictive-technology-and-the-business-of-keeping-us-hooked/, March 24, 2017.

Alter, Adam. "Why Our Screens Make Us Less Happy" *TED Conference*. https://www.ted.com/talks/adam_alter_why_our_screens_make_us_less_happy, April 2017.

Anderson, Jill. “Smartphones, teens, and unhappiness” *The Harvard Gazette, Harvard University.* https://news.harvard.edu/gazette/story/2018/06/gse-phones-study/, June 20, 2018.

Anderson, Monica and Jingjing Jiang. “Teens, Social Media & Technology 2018” *Pew Research Center*. https://www.pewresearch.org/internet/wp-content/uploads/sites/9/2018/05/PI_2018.05.31_TeensTech_FINAL.pdf, May 31, 2018.

Anderson, Porter. “Study: Multitasking is counterproductive” *CNN*. https://www.cnn.com/2001/CAREER/trends/08/05/multitasking.study/index.html

Andersson, Hilary. “Social media apps are ‘deliberately’ addictive to users” *BBC.* https://www.bbc.com/news/technology-44640959, July 4, 2018.

Arum, Richard. “College Graduates: Satisfied, but Adrift” in *The State of the American Mind*. p. 73, Mark Bauerlein and Adam Bellow, eds. West Conshohocken, PA: Templeton, 2015.

Askvik, Eva Ose, PhD; F. R. (Ruud) van der Weel, PhD; Audrey L. H. van der Meer, PhD. “The Importance of Cursive Handwriting Over Typewriting for Learning in the Classroom: A High-Density EEG Study of 12-Year-Old Children and Young Adults.” *Frontiers in Psychology*.
https://doi.org/10.3389/fpsyg.2020.01810
https://www.frontiersin.org/articles/10.3389/fpsyg.2020.01810/full, July 28, 2020.

Atherton, Marc. “Silicon Valley’s Secret Sauce” *Swipe Left For Addiction*. https://swipeleftforaddiction.com/2018/05/30/silicon-valleys-secret-sauce/, accessed July 28, 2018.
Updated source site:
“The Attention Economy, Internet Addiction and Unintended Consequences” *Swipe Left For Addiction.*
https://swipeleftforaddiction.wordpress.com/2018/05/11/the-attention-economy-internet-addiction-and-unintended-consequences/, May 11, 2018.

Atir, Stav; Kristina A. Wald; Nicholas Epley. "Talking with strangers is surprisingly informative" *Proceedings of the National Academy of Sciences*. Vol. 119, No. 34.
https://www.pnas.org/doi/10.1073/pnas.2206992119, August 16, 2022.

Bailey, Blake. *Philip Roth: The Biography*. (W. W. Norton & Company; 1st Edition [Hardcover], 2021).

Baron, Naomi S. *How We Read Now: Strategic Choices for Print, Screen, and Audio*. New York: Oxford University Press. March 22, 2021.

Baron, Naomi S. "Why do we remember more by reading in print vs. on a screen?" *Neuropsych*.
https://bigthink.com/neuropsych/reading-memory/, May 9, 2021.

Becker-Phelps, Leslie. "Don't Just React: Choose Your Response" *Psychology Today*.
https://www.psychologytoday.com/blog/making-change/201307/dont-just-react-choose-your-response, July 23, 2013.

Beland, Louis-Philippe, and Richard Murphy. "Ill Communication: Technology, distraction & student performance" *Labour Economics*. Elsevier B.V.
https://www.sciencedirect.com/science/article/abs/pii/S0927537116300136, April 16, 2016

Belkhir, Lotfi and Ahmed Elmeligi. "Assessing ICT global emissions footprint: Trends to 2040 & recommendations" *Journal of Cleaner Production*, Volume 177, Pages 448-463. Elsevier B.V.
https://doi.org/10.1016/j.jclepro.2017.12.239, March 10, 2018.

Belkin, Douglas; Ben Chapma; and Ben Kesling. "'How Do I Do That?' The New Hires of 2023 Are Unprepared for Work" *The Wall Street Journal*.
https://www.wsj.com/articles/lost-learning-remote-pandemic-workplace-skills-new-employees-51351b33.

Berger, Eric. “By 2010 most science Ph.D.s will go to foreign-born students” *SciGuy*. http://blog.chron.com/sciguy/2007/11/by-2010-most-science-ph-d-s-will-go-to-foreign-born-students/, November 21, 2007.

Berman, Robby. “A home library can have a powerful effect on children: A new study finds that simply growing up in a home with enough books increases adult literacy and math prowess.” *BigThink*. https://bigthink.com/the-present/mind-brain-home-library-benefits, October 13, 2018.

Bernstein, Emily E. and Richard J. McNally. “Acute aerobic exercise helps overcome emotion regulation deficits” *Cognition and Emotion*, DOI: 10.1080/02699931.2016.1168284 http://dx.doi.org/10.1080/02699931.2016.1168284, April 4, 2016.

Birdsong, Kristina. “This is Your Child’s Brain on TV” *Scientific Learning*. http://54.186.226.228/blog/your-childs-brain-tv, March 22, 2016.

Birken, Emily Guy. “This Is How Americans Spent Their Money in the 1950s” *WiseBread.* https://www.wisebread.com/this-is-how-americans-spent-their-money-in-the-1950s, January 12, 2016.

Blachowicz, James. “There Is No Scientific Method” *The Stone, New York Times*. http://www.nytimes.com/2016/07/04/opinion/there-is-no-scientific-method.html, July 4, 2016.

Bowles, Nellie. “Human Contact Is Now a Luxury Good: Screens used to be for the elite. Now avoiding them is a status symbol.” *The New York Times*. https://www.nytimes.com/2019/03/23/sunday-review/human-contact-luxury-screens.html, March 23, 2019.

Bowles, Nellie. “The Digital Gap Between Rich and Poor Kids Is Not What We Expected” *The New York Times*. https://www.nytimes.com/2018/10/26/style/digital-divide-screens-schools.html, October 26, 2018.

Boyse, Kyla. "Television and Children" *Michigan Medicine* (University of Michigan). http://www.med.umich.edu/yourchild/topics/tv.htm, August 2010. Updated: Radesky, Jenny MD. "Kids and Digital Media" Michigan Medicine (University of Michigan Health). https://www.mottchildren.org/posts/your-child/kids-and-digital-media, May 2017.

Bradberry, Travis and Jean Greaves. *Emotional Intelligence 2.0* (San Diego, California: TalentSmart, 2009).

- Bradberry, Travis. "9 Habits of Profoundly Influential People". https://www.linkedin.com/pulse/critical-habits-profoundly-influential-people-dr-travis-bradberry/, July 20, 2015.
- Bradberry, Travis. "9 Surprising Things Ultra Productive People Do Every Day". https://www.linkedin.com/pulse/surprising-things-ultra-productive-people-do-every-day-bradberry/, November 7, 2016.
- Bradberry, Travis. "11 Things Ultra-Productive People Do Differently" *Forbes*. https://www.forbes.com/sites/travisbradberry/2015/05/13/11-things-ultra-productive-people-do-differently/, May 13, 2015.
- Bradberry, Travis. "12 Habits of Genuine People". https://www.linkedin.com/pulse/importance-being-genuine-dr-travis-bradberry/, November 15, 2015.
- Bradberry, Travis. "13 Habits of Exceptionally Likeable People". https://www.linkedin.com/pulse/13-habits-exceptionally-likeable-people-dr-travis-bradberry, January 27, 2015.
- Bradberry, Travis. "13 Things Mentally Strong People Won't Do". https://www.linkedin.com/pulse/13-things-mentally-strong-people-wont-do-dr-travis-bradberry/, September 11, 2017.
- Bradberry, Travis. "Eight Habits of Considerate People". https://www.linkedin.com/pulse/eight-habits-considerate-people-dr-travis-bradberry, November 8, 2017.
- Bradberry, Travis. "These are the habits that mentally strong people rely on" *World Economic Forum*. https://www.weforum.org/agenda/2016/10/habits-to-help-you-develop-mental-strength, October 26, 2016.

Brandslet, Steinar. "Students learn better writing by hand: We learn much better if we take notes the old-fashioned way, by hand, and now researchers know more about why this is." *Futurity*. Original Study DOI: 10.3389/fpsyg.2023.1219945. https://www.futurity.org/writing-by-hand-learning-3253352-2/, October 15, 2024.

Brooks, David. "Why Your Social Life Is Not What It Should Be" *The New York Times.* https://www.nytimes.com/2022/08/25/opinion/social-life-talk-strangers.html, August 25, 2022.

Brown, Stuart. "Play, Spirit, and Character" *Interview for* On Being with Krista Tippett*, a Peabody Award-winning public radio show and podcast.* https://onbeing.org/programs/stuart-brown-play-spirit-and-character/, June 19, 2014.

Bryson, Carey. "Babies and TV: Is Screen Time Good for Your Little One?" *ThoughtCo.* https://www.thoughtco.com/should-babies-watch-tv-2107982, June 22, 2017.

Buckner, Candace. "NBA players know they're addicted to their phones. Good luck getting them to unplug" *The Washington Post.* https://www.washingtonpost.com/sports/nba-players-know-theyre-addicted-to-their-phones-good-luck-getting-them-to-unplug/2018/03/19/6165cb96-2563-11e8-b79d-f3d931db7f68_story.html, March 19, 2018.

Burhan, R., & Moradzadeh, J. "Neurotransmitter Dopamine (DA) and its Role in the Development of Social Media Addiction" *Journal of Neurology, 11*(7), 507. https://www.iomcworld.org/open-access/neurotransmitter-dopamine-da-and-its-role-in-the-development-of-social-media-addiction-59222.html, 2020.

Carey, Kevin. "Americans Think We Have the World's Best Colleges. We Don't." *The New York Times.* https://www.nytimes.com/2014/06/29/upshot/americans-think-we-have-the-worlds-best-colleges-we-dont.html, June 28, 2014.

Carter, Christine. “Three Risks of Too Much Screen Time for Teens” *Greater Good.* Greater Good Science Center (GGSC), University of California, Berkeley. https://greatergood.berkeley.edu/article/item/three_risks_of_too_much_screen_time_for_teens, November 27, 2018.

Chervinsky, Lindsay M. “George Washington: Life in Brief” Miller Center, University of Virginia.
https://millercenter.org/president/washington, Accessed March 18, 2022
https://millercenter.org/president/washington/life-in-brief, Accessed March 18, 2022.

“Children who watch ‘excessive’ amounts of TV are more likely to have criminal convictions, exhibit aggression and experience negative emotions: study.” *New York Daily News*.
http://www.nydailynews.com/life-style/health/kids-watch-excessive-tv-criminal-convictions-young-adulthood-study-article-1.1267868, February 19, 2013.

Christensen, Allan Stromfeldt. “Lemminged: to be herded off the peak oil cliff by filmmakers” *TransitionVoice.com*.
http://transitionvoice.com/2014/11/lemminged-to-be-herded-off-the-peak-oil-cliff-by-filmmakers/, November 11, 2014.

Chua, Celestine. “Top 10 Reasons You Should Stop Watching TV” *Personal Excellence: Be Your Best Self, Live Your Best Life.*
http://personalexcellence.co/blog/top-10-reasons-you-should-stop-watching-tv/, May 2, 2010

Cicero, Marcus Tullius. *De Oratore,* Book II §36. The Loeb Classical Library, English translation Cambridge, Massachusetts, Harvard University Press, 1967. https://archive.org/stream/cicerodeoratore01ciceuoft/cicerodeoratore01ciceuoft_djvu.txt, 55 BC.

Constine, Josh. “Jeff Bezos’ guide to life” *TechCrunch.*
https://techcrunch.com/2017/11/05/jeff-bezos-guide-to-life/, November 5, 2017.

Cooper, Anderson. "Groundbreaking study examines effects of screen time on kids" *CBSNews*. https://www.cbsnews.com/news/groundbreaking-study-examines-effects-of-screen-time-on-kids-60-minutes/, December 9, 2018.

Cooper, Belle Beth. "How to Be a Success at Everything" *Fast Company*. https://www.fastcompany.com/3056613/how-i-became-a-morning-person-read-more-books-and-learned-, February 12, 2016.

Cross, Daile. "Treat TV like passive smoking when it comes to children" *WA Today*. https://www.watoday.com.au/national/western-australia/treat-tv-like-passive-smoking-when-it-comes-to-children-20171106-gzftv4.html, November 9, 2017.

Courogen, Carrie. "9 Harsh Realities About Graduating College That I Wish Someone Had Warned Me About" *Bustle*. https://www.bustle.com/articles/80461-9-harsh-realities-about-graduating-college-that-i-wish-someone-had-warned-me-about, May 4, 2015.

Cuddy, Alice. "The IQ of Europeans is dropping due to technology, say researchers" *Euronews*. https://www.euronews.com/2017/12/29/the-iq-of-europeans-is-dropping-due-to-technology-say-researchers, December 29, 2017.

Curry, David. "BeReal Revenue and Usage Statistics (2022)" *Business of Apps*. https://www.businessofapps.com/data/bereal-statistics/, November 10, 2022.

Dahl, Melissa. "How Neuroscientists Explain the Mind-Clearing Magic of Running" Science of Us, *Huffington Post*. https://www.huffingtonpost.com/science-of-us/how-neuroscientists-expla_b_9787466.html, December 6, 2017.

Deese, Kaelan. "Oregon governor signs bill ending reading and math proficiency requirements for graduation" *YahooNews*. https://www.yahoo.com/now/oregon-governor-signs-bill-ending-154100667.html, August 10, 2021.

Dong, Guangheng; Yanbo Hu; Xiao Lin. "Reward/punishment sensitivities among internet addicts: Implications for their addictive behaviors" *Progress in Neuro-Psychopharmacology and Biological Psychiatry*, Volume 46, October 1, 2013, Pages 139-145. https://www.sciencedirect.com/science/article/pii/S0278584613001486, July 19, 2013.

Doverspike, William F., Ph.D. "How To Make Yourself Miserable: Discovering the Secrets to Unhappiness" *Georgia Psychological Association*. Atlanta. http://gapsychology.org/displaycommon.cfm?an=1&subarticlenbr=341, accessed September 29, 2014.
Also: Drdoverspike.com. http://drwilliamdoverspike.com/files/how_to_make_yourself_miserable_revised_version.pdf.

Dovey, Ceridwen. "Can Reading Make You Happier?" *The New Yorker*. http://www.newyorker.com/culture/cultural-comment/can-reading-make-you-happier, June 9, 2015.

Dunckley, Victoria L., MD. "Dumb & Dumber: Interactive Screentime is Worse than TV" *Psychology Today.* https://www.psychologytoday.com/us/blog/mental-wealth/201408/dumb-dumber-interactive-screentime-is-worse-tv, August 29, 2014.

Dunckley, Victoria L., MD. "Electronic Screen Syndrome: An Unrecognized Disorder? Screentime and the rise of mental disorders in children." *Psychology Today*. https://www.psychologytoday.com/us/blog/mental-wealth/201207/electronic-screen-syndrome-unrecognized-disorder, July 23, 2012.

Dunckley, Victoria L., MD. "Gray Matters: Too Much Screen Time Damages the Brain" *Psychology Today.* https://www.psychologytoday.com/us/blog/mental-wealth/201402/gray-matters-too-much-screen-time-damages-the-brain, February 27, 2014.

Dunckley, Victoria L., MD. "Screentime and Arrested Social Development" *Psychology Today.* https://www.psychologytoday.com/us/blog/mental-wealth/201606/screentime-and-arrested-social-development, June 30, 2016.

Dunckley, Victoria L., MD. Video Talk (*YouTube*). https://youtu.be/SWQuNnCdN5Y, November 8, 2018.

Dunning, David. "We Are All Confident Idiots" *Pacific Standard.* https://www3.nd.edu/~ghaeffel/ConfidentIdiots.pdf, October 27, 2014.

Dunning, David and Justin Kruger. "Unskilled and Unaware of It: How Difficulties in Recognizing One's Own Incompetence Lead to Inflated Self-Assessments" *Journal of Personality and Social Psychology.* 1999, Vol. 77, No. 6, 1121-1134. American Psychological Association. https://psycnet.apa.org/doiLanding?doi=10.1037%2F0022-3514.77.6.1121

Dwyer, Ryan J., Kostadin Kushlev, Elizabeth W. Dunn. "Smartphone use undermines enjoyment of face-to-face social interactions" *Journal of Experimental Social Psychology.* Elsevier B.V. https://www.sciencedirect.com/science/article/pii/S0022103117301737, September 2018.

Educational Testing Service (ETS). "America's Skills Challenge: Millennials and the Future" *The ETS Center for Research on Human Capital and Education.* Princeton, NJ: January 2015. https://www.ets.org/s/research/30079/asc-millennials-and-the-future.pdf.

Epley, Nicholas; Juliana Schroeder. "Mistakenly Seeking Solitude" *Journal of Experimental Psycholog*y, American Psychological Association. 2014, Vol. 143, No. 5, pp. 1980–1999.
https://uploads-ssl.webflow.com/5c484e0f4aa6f839dc553c45/5c9540a7e291048443c0c276_EpleySchroederJEPG2014.pdf
https://doi.org/10.1037/a0037323, accessed August 29, 2022.

Epley, Nicholas; Michael Kardas; Xuan Zhao; Stav Atir; Juliana Schroeder. "Undersociality: miscalibrated social cognition can inhibit social connection" *Trends in Cognitive Sciences*. Vol. 26, Issue 5, pp. 406–418, May 1, 2022. https://doi.org/10.1016/j.tics.2022.02.007, March 24, 2022.

Fabritius, Friederike; Hans W. Hagemann. *The Leading Brain: Neuroscience Hacks to Work Smarter, Better, Happier*. TarcherPerigee, February 20, 2018.

Fang, Richard. The psychology of why social media is so addictive. *UX Collective*.
https://uxdesign.cc/the-psychology-of-why-social-media-is-so-addictive-67830266657d, September 21, 2020.

Fields, Douglas. "Watching TV Alters Children's Brain Structure and Lowers IQ". http://rdouglasfields.com/2015/05/watching-tv-alters-childrens-brain-structure-and-lowers-iq/, May 4, 2015.

Fisher, Matthew et al. "Searching for Explanations: How the Internet Inflates Estimates of Internal Knowledge" *Journal of Experimental Psychology* 144(3). pp. 674–687, June 2015.

Flynn, James and Michael Shayer. "IQ decline and Piaget: Does the rot start at the top?" *Intelligence,* Volume 66, Pages 112-121. Elsevier B.V. https://doi.org/10.1016/j.intell.2017.11.010, December 8, 2017.

Fox, EJ and Mike Spies. "Who Was America's Most Well-Spoken President?" *Vocativ*.
http://www.vocativ.com/interactive/usa/us-politics/presidential-readability/, October 10, 2014.
and
Fairchild, Quentin B. "Presidential speechmaking in an age of ignorance" News-Press.
https://www.news-press.com/story/opinion/contributors/2014/11/05/pr

Freed, Richard. "The Tech Industry's War on Kids: How psychology is being used as a weapon against children" Blog, Child and adolescent psychologist Richard Freed.
https://medium.com/@richardnfreed/the-tech-industrys-psychological-war-on-kids-c452870464ce, March 12, 2018.

Frey, Bruno S.; Christine Benesch; Alois Stutzer. “Does watching TV make us happy?” *Journal of Economic Psychology*, Volume 28, Issue 3, Pages 283–313, June 2007. Elsevier B.V. (ScienceDirect.com). https://www.bsfrey.ch/wp-content/uploads/2021/08/does-watching-tv-make-us-happy.pdf, February 14, 2007.
Updated URLs:
https://doi.org/10.1016/j.joep.2007.02.001,
https://www.sciencedirect.com/science/article/abs/pii/S0167487007000104.

Friedman, Thomas. *Thank You for Being Late.* New York: Farrar, Straus and Giroux (Picador), 2017.

Frierson, William. “Dream vs. Reality: What Happens After Graduation” *College Recruiter*. https://www.collegerecruiter.com/blog/2015/07/14/dream-vs-reality-what-happens-after-graduation/, July 14, 2015.

Gardner, Amanda. “TV watching raises risk of health problems, dying young” *CNN*. http://www.cnn.com/2011/HEALTH/06/14/tv.watching.unhealthy/, June 14, 2011.

Anderson-Niles, Amanda. “Gender Switch: When Did Men Become So Feminine?” Urban Belle Magazine. https://urbanbellemag.com/2010/10/gender-switch-when-did-men-become-so-feminine/, October 18, 2010.

Goldsmith, Belinda. “Watching hours of TV daily could shorten your life – study” Ed. Miral Fahmy. *Reuters*. http://in.reuters.com/article/worldNews/idINIndia-45316820100111, January 12, 2010.

Gollwitzer, Peter M., et al., “When Intentions Go Public: Does Social Reality Widen the Intention-Behavior Gap?” *Psychological Science* 20, no. 5 (May 1, 2009): 612. http://journals.sagepub.com/doi/abs/10.1111/j.1467-9280.2009.02336.x, May 1, 2009.

Grabmeier, Jeff. "Both liberals, conservatives can have science bias: Study finds different topics bedevil the left and right" *The Ohio State University*. https://news.osu.edu/news/2015/02/09/both-liberals-conservatives-can-have-science-bias/, February 09, 2015.

Ha, Thu-Huong. "New research links reading books with longer life" *Quartz*. http://qz.com/754109/new-research-links-reading-books-with-longer-life/, August 10, 2016.

Haden, Jeff. "How the Smartest Minds Use Intuition and Emotional Intelligence to Make Better Decisions, Backed by Science: Experience matters, but research shows high emotional intelligence can also make intuitive decisions even more accurate" *Inc.com*. https://www.inc.com/jeff-haden/how-smartest-minds-use-intuition-emotional-intelligence-to-make-better-decisions-backed-by-science.html, April 8, 2022.

Haden, Jeff. *The Motivation Myth.* New York: Penguin, 2018.

Hamilton, Jon. "How Play Wires Kids' Brains For Social and Academic Success" *KQED News.* KQED.org, National Public Radio (NPR), Copyright 2014. http://ww2.kqed.org/mindshift/2014/08/07/how-play-wires-kids-brains-for-social-and-academic-success/, August 7, 2014.

Han, Seunghee; Jang Hyun Kim; Ki Joon Kim. "Understanding Nomophobia: Structural Equation Modeling and Semantic Network Analysis of Smartphone Separation Anxiety." *US National Library of Medicine National Institutes of Health.* https://www.ncbi.nlm.nih.gov/pubmed/28650222, June 26, 2017.

Hatano, Aya, PhD, Kyoto University; Cansu Ogulmus, PhD, University of Tübingen; Kou Murayama, PhD, University of Tübingen; Hiroaki Shigemasu, PhD, Kochi University of Technology. "Thinking About Thinking: People Underestimate How Enjoyable and Engaging Just Waiting Is" *Journal of Experimental Psychology: General*. https://www.apa.org/news/press/releases/2022/07/thoughts-mind-wander, July 28, 2022.

Hawkley, Louise C., Ph.D. ; John T. Cacioppo, Ph.D. "Loneliness Matters: A Theoretical and Empirical Review of Consequences and Mechanisms" *Annals of Behavioral Medicine*, Volume 40, Issue 2, October 2010, Pages 218–227. https://doi.org/10.1007/s12160-010-9210-8.
Healey, Aleeya and Alan Mendelsohn. "Selecting Appropriate Toys for Young Children in the Digital Era" *American Academy of Pediatrics*. http://pediatrics.aappublications.org/content/early/2018/11/29/peds.2018-3348, December 2018.
Print Source: Pediatrics, Volume 143, number 1, January 2019 (Copyright © American Academy of Pediatrics 2019).

Healy, Jane M. *Endangered Minds: Why Children Don't Think And What We Can Do About It*. Simon & Schuster. September 1, 1999. Jane M. Healy. "Endangered Minds" Creating the Future: Perspectives on Educational Change. Ed. Dee Dickinson. Johns Hopkins University. 2012.

Heilbrunn Timeline of Art History. The Metropolitan Museum of Art. http://www.metmuseum.org/toah/hd/gogh/hd_gogh.htm, July, 30 2015.

Heitman, Danny. "I'm Revisiting the Books of My Youth" *The Wall Street Journal*, Vol. CCLXXXI No. 97. Thursday April 27, 2023.

Henderson, Emma. "Watching lots of TV 'makes you stupid'" *The Independent.* http://www.independent.co.uk/news/science/watching-lots-of-tv-makes-you-stupid-says-american-universities-a6759026.html, December 3, 2015.

Heshmat, Shahram. "What Is Confirmation Bias?" *Psychology Today.* https://www.psychologytoday.com/blog/science-choice/201504/what-is-confirmation-bias, April 23, 2015.

Hickman, Martin. "Watching TV 'makes toddlers less intelligent'" *The Independent.* http://www.independent.co.uk/news/education/education-news/watching-tv-makes-toddlers-less-intelligent-1960856.html, May 2, 2010.

Higgins, E. Tory. "Self-Discrepancy Theory" *Advances in Experimental Social Psychology*, Volume 22. Leonard Berkowitz, ed. (Cambridge, Massachusetts: Academic Press. March 28, 1989) pp. 93–135 (specific referenced quotes on pp. 118–119).

Hill, David L. "Why to Avoid TV Before Age 2" (early brain development). *American Academy of Pediatrics*. http://www.healthychildren.org/English/family-life/Media/Pages/Why-to-Avoid-TV-Before-Age-2.aspx, May 11, 2013.
Original Source: Dad to Dad: Parenting Like a Pro (Copyright © American Academy of Pediatrics 2012).

Hinckley, David. "Average American watches 5 hours of TV per day, report shows" *New York Daily News*. https://www.nydailynews.com/life-style/average-american-watches-5-hours-tv-day-article-1.1711954, March 5, 2014.

Hoang, Tina D., MSPH; Jared Reis, PhD; Na Zhu, MD, MPH; David R. Jacobs Jr, PhD; Lenore J. Launer, PhD; Rachel A. Whitmer, PhD; Stephen Sidney, MD; Kristine Yaffe, MD. "Effect of Early Adult Patterns of Physical Activity and Television Viewing on Midlife Cognitive Function" *JAMA Psychiatry* (The Journal of the American Medical Association). http://archpsyc.jamanetwork.com/article.aspx?articleid=2471270, January 2016, Vol 73, No. 1.

Holthaus, Eric and Chris Kirk. "A Filthy History: Interactive map: Which countries have emitted the most carbon since 1850?" *Slate.com*. http://www.slate.com/articles/technology/future_tense/2014/05/carbon_dioxide_emissions_by_country_over_time_the_worst_global_warming_polluters.html.

"HOME-SCHOOLING: Outstanding results on national tests" *The Washington Times*. http://www.washingtontimes.com/news/2009/aug/30/home-schooling-outstanding-results-national-tests/, August 30, 2009.

"How Satellites Work With Mobile Phones" *CompareMyMobile*. http://blog.comparemymobile.com/how-satellites-work-with-mobile-phones/, accessed December 21, 2017.

Hu, Elise. "Facebook Makes Us Sadder And Less Satisfied, Study Finds" *National Public Radio*. http://www.npr.org/blogs/alltechconsidered/2013/08/19/213568763/researchers-facebook-makes-us-sadder-and-less-satisfied, August 20, 2013.

Hunt, Melissa G.; Rachel Marx; Courtney Lipson; Jordyn Young. "No More FOMO: Limiting Social Media Decreases Loneliness and Depression" *Journal of Social and Clinical Psychology*. Guilford Press. https://guilfordjournals.com/doi/10.1521/jscp.2018.37.10.751, October 18, 2018.

Hurley, Bevan. "Family claim 10-year-old obsessed with gadgets fatally shot his mother for refusing to buy him a VR headset" *The Independent*. https://www.independent.co.uk/news/world/americas/crime/quiana-mann-son-murder-virtual-reality-b2253013.html, December 29, 2022.

Hutton, John S., MS, MD; Jonathan Dudley, PhD; Tzipi Horowitz-Kraus, PhD. "Associations Between Screen-Based Media Use and Brain White Matter Integrity in Preschool-Aged Children" *American Medical Association: JAMA Pediatr. 2020;174(1):e193869. doi:10.1001/jamapediatrics.2019.3869*. https://jamanetwork.com/journals/jamapediatrics/fullarticle/2754101, November 4, 2019.

Hyodo, Kazuki and Ippeita Dan, Yasushi Kyutoku, Kazuya Suwabe, Kyeongho Byun, Genta Ochi, Morimasa Kato, Hideaki Soya. "The association between aerobic fitness and cognitive function in older men mediated by frontal lateralization" *NeuroImage*, v. 125, pp. 291–300, January 15, 2016.

"I'm Addicted to Television": The Personality, Imagination, and TV Watching Patterns of Self-Identified TV Addicts. Robert D. McIlwraith in *Journal of Broadcasting and Electronic Media*, Vol. 42, No. 3, pages 371--386; Summer 1998.

Ingraham, Christopher. "Today's men are not nearly as strong as their dads were, researchers say" *Washington Post*. https://www.washingtonpost.com/news/wonk/wp/2016/08/15/todays-men-are-nowhere-near-as-strong-as-their-dads-were-researchers-say/, August 15, 2016.

Iqbal, Mansoor. "TikTok Revenue and Usage Statistics (2022)" *Business of Apps*. https://www.businessofapps.com/data/tik-tok-statistics/, November 11, 2022.

Iqbal, Mansoor. "Snapchat Revenue and Usage Statistics (2022)" *Business of Apps*.
https://www.businessofapps.com/data/snapchat-statistics/, August 31, 2022.

Iqbal, Mansoor. "Facebook Revenue and Usage Statistics (2022)" *Business of Apps*.
https://www.businessofapps.com/data/facebook-statistics/, November 24, 2022.

Iqbal, Mansoor. "Instagram Revenue and Usage Statistics (2022)" *Business of Apps*.
https://www.businessofapps.com/data/instagram-statistics/, September 6, 2022.

Iqbal, Mansoor. "Twitter Revenue and Usage Statistics (2022)" *Business of Apps*.
https://www.businessofapps.com/data/twitter-statistics/, November 4, 2022.

Jacobs, Tom. "Searching the Internet Creates an Illusion of Knowledge" *Pacific Standard online.*
https://psmag.com/environment/searching-internet-creates-the-illusion-of-knowledge-, April 1, 2015.

James, Henry. *The Tragic Muse.* p. 582. London: Rupert Hart-Davis, 1948.

Jaschik, Scott. "Let the Right Ones In" *Slate.com*.
http://www.slate.com/articles/life/inside_higher_ed/2014/10/college_admissions_rose_hulman_institute_of_technology_uses_locus_of_control.html, October 30, 2014.

Jayson, Sharon. "Generation Y's goal? Wealth and fame" *USA Today*.
http://usatoday30.usatoday.com/news/nation/2007-01-09-gen-y-cover_x.htm, Posted January 9,2007, Updated January 10, 2007.

Johnson, Stephen. "China's schoolkids beat American students in all academic categories" *bigthink.com*. https://bigthink.com/the-present/pisa-test-china/#rebelltitem4, December 5, 2019.

Jordan, Alexander H., Benoît Monin, Carol S. Dweck, Benjamin J. Lovett, Oliver P. John, and James J. Gross. "Misery Has More Company Than People Think: Underestimating the Prevalence of Others' Negative Emotions". *Personality and Social Psychology Bulletin* January 2011 37: 120-135. National Institutes of Health. http://www.ncbi.nlm.nih.gov/pmc/articles/PMC4138214/, January 2011, latest update August 19, 2014.

Joyce, Amy. "How helicopter parents are ruining college students" *The Washington Post*. http://www.washingtonpost.com/news/parenting/wp/2014/09/02/how-helicopter-parents-are-ruining-college-students/, September 2, 2014

Kardaras, Nicholas. "It's 'digital heroin': How screens turn kids into psychotic junkies" *The New York Post*. https://nypost.com/2016/08/27/its-digital-heroin-how-screens-turn-kids-into-psychotic-junkies/, August 27, 2016.

Kardaras, Nicholas. "Kids turn violent as parents battle 'digital heroin' addiction" *The New York Post*. https://nypost.com/2016/08/27/its-digital-heroin-how-screens-turn-kids-into-psychotic-junkies/, December 17, 2016.

Kardaras, Nicholas. "The case against screens in schools" *The New York Post*. https://nypost.com/2016/12/17/the-case-against-screens-in-schools/, December 17, 2016.

Kardas, Michael; Amit Kumar; Nicholas Epley. "Overly Shallow?: Miscalibrated Expectations Create a Barrier to Deeper Conversation" *Personality and Social Psychology: Attitudes and Social Cognition*. American Psychological Association. https://www.apa.org/pubs/journals/releases/psp-pspa0000281.pdf, May 4, 2021.

Kardas, Michael; J. Schroeder; and E. O'Brien. "Keep talking: (Mis)understanding the hedonic trajectory of conversation" *Journal of Personality and Social Psychology*. Advance online publication. https://psycnet.apa.org/record/2022-16089-001?doi=1 https://psycnet.apa.org/doiLanding?doi=10.1037%2Fpspi0000379, accessed August 29, 2022.

Kennedy-Moore, Eileen. "Do Boys Need Rough and Tumble Play?" *Psychology Today*. https://www.psychologytoday.com/blog/growing-friendships/201506/do-boys-need-rough-and-tumble-play, June 30, 2015.

Keohane, Joe. "How Facts Backfire: Researchers Discover a Surprising Threat to Democracy: Our Brains" Boston Globe online. http://archive.boston.com/news/science/articles/2010/07/11/how_facts_backfire/, July 11, 2010.

Kharpal, Arjun; Lauren Feiner. "China to ban kids from playing online games for more than three hours per week" *CNBC*. https://cnb.cx/2V4tkMw (also https://www.cnbc.com/2021/08/30/china-to-ban-kids-from-playing-online-games-for-more-than-three-hours-per-week.html), August 30 2021.

Kim, Juliana. "How 'goblin mode' became Oxford's word of the year" *NPR*. https://www.npr.org/2022/12/05/1140696560/oxford-word-2022-goblin-mode, December 5, 2022.

Kim, Kyung Hee. "The Creativity Crisis: The Decrease in Creative Thinking Scores on the Torrance Tests of Creative Thinking" *Creativity Research Journal*, 23:4, 285–295. Routledge, Taylor & Francis Group. https://www.tandfonline.com/doi/abs/10.1080/10400419.2011.627805, November 9, 2011.
Downloaded by the College of William and Mary: https://kkim.wmwikis.net/file/view/Kim_2011_Creativity_crisis.pdf, April 3, 2012.

Kirk, Chris. "Five Charts That Show Americans Families' Debt Crisis" *Slate.com.* (Data from the Federal Reserve Bank, St. Louis.) http://www.slate.com/articles/business/the_united_states_of_debt/2016/05/the_rise_of_household_debt_in_the_u_s_in_five_charts.html, May 12, 2016.

Klass, Perri, M.D. "Why Handwriting Is Still Essential in the Keyboard Age" *The New York Times.* https://well.blogs.nytimes.com/2016/06/20/why-handwriting-is-still-essential-in-the-keyboard-age, June 20, 2016.

Knott, Stephen, ed. "Life Before the Presidency" *American President: George Washington (1732–1799), Essays on George Washington and His Administration*. Miller Center, University of Virginia. http://millercenter.org/president/washington/essays/biography/2, accessed November 19, 2014.

Koetsier, John. "Digital Crack Cocaine: The Science Behind TikTok's Success" *Forbes*. https://www.forbes.com/sites/johnkoetsier/2020/01/18/digital-crack-cocaine-the-science-behind-tiktoks-success/, January 18, 2020.

Kopf, Dan. "The share of American young adults living with their parents is the highest in 75 years" *Quartz.* https://qz.com/1248081/the-share-of-americans-age-25-29-living-with-parents-is-the-highest-in-75-years/, April 10, 2018.

Krieger, Richard Alan. *Civilization's Quotations: Life's Ideal.* (New York: Algora Publishing, 2002) p. 122.

Kross, Ethan; Philippe Verduyn, Emre Demiralp, Jiyoung Park, David Seungjae Lee, Natalie Lin, Holly Shablack, John Jonides, Oscar Ybarra. "Facebook Use Predicts Declines in Subjective Well-Being in Young Adults" *PLoS ONE* 8(8): e69841. doi:10.1371/journal.pone.0069841. http://www.plosone.org/article/info%3Adoi%2F10.1371%2Fjournal.pone.0069841, August 14, 2013.

Krugman, Herbert E. "Brain Wave Measures of Media Involvement" *How Advertising Works: The Role of Research.* (New York SAGE Publications, 1998) pp. 139–151.

Krugman, Herbert E., and Eugene L. Hartley. "Passive Learning From Television" *The Public Opinion Quarterly,* Vol. 34, No. 2. (Oxford University Press, 1970) pp. 184-190.

Kubey, Robert and Mihaly Csikszentmihalyi. "Television Addiction is no mere metaphor" *Scientific American*. http://www.academia.edu/5065840/Television_Addiction_is_no_mere_metaphor, accessed January 23, 2016.

Kubey, Robert and Mihaly Csikszentmihalyi. *Television and the Quality of Life: How Viewing Shapes Everyday Experience.* Lawrence Erlbaum Associates, 1990.

Kubey, Robert; Michael J. Lavin and John R. Barrows. "Internet Use and Collegiate Academic Performance Decrements: Early Findings". *Journal of Communication*, Vol. 51, No. 2, pages 366--382; June 2001.

Kubu, Cynthia, PhD; and Andre Machado, MD. "The Science Is Clear: Why Multitasking Doesn't Work" *Cleveland Clinic Health Essentials*. https://health.clevelandclinic.org/science-clear-multitasking-doesnt-work/, June 1, 2017, updated March 10, 2021.

LaMotte, Sandee. "MRIs show screen time linked to lower brain development in preschoolers" *CNN*. https://www.cnn.com/2019/11/04/health/screen-time-lower-brain-development-preschoolers-wellness/index.html, November 4, 2019.

Lang, Annie. The Limited Capacity Model of Mediated Message Processing. *Journal of Communication*, Vol. 50, No. 1, pages 46--70; March 2000.

Lee-Won, R. J., Herzog, L., & Park, S. G. "Hooked on Facebook: The role of social anxiety and need for social assurance in problematic use of Facebook" *Cyberpsychology, Behavior, and Social Networking, 18*(10), 567-574. https://pubmed.ncbi.nlm.nih.gov/26383178/, Sep 18, 2015.

Leonard, Tom. "'Passive' TV Can Harm Your Baby's Speech Making It Harder for Them to Later Cope In School" *Daily Mail*. http://www.dailymail.co.uk/news/article-2054950/Passive-TV-watching-harm-babies-speech.html, October 28, 2011.

Liebert, Mary Ann. "Understanding smartphone separation anxiety and what smartphones mean to people". *Phys.org.* https://phys.org/news/2017-08-smartphone-anxiety-smartphones-people.html, August 14, 2017.

Lin, L. Y., Sidani, J. E., Shensa, A., Radovic, A., Miller, E., Colditz, J. B., Primack, B. A. "Association between social media use and depression among U.S. young adults" *Depression and anxiety*, *33*(4), 323-331. https://doi.org/10.1002/da.22466, January 19, 2016.

Lissak, Gadi. "Adverse physiological and psychological effects of screen time on children and adolescents: Literature review and case study". Environmental Research, Volume 164, July 2018, Pages 149-157. https://doi.org/10.1016/j.envres.2018.01.015 (also https://www.sciencedirect.com/science/article/pii/S001393511830015X), February, 27 2018.

MacBeth, Tannis M., ed. Television Dependence, Diagnosis, and Prevention. Robert W. Kubey in Tuning in to Young Viewers: Social Science Perspectives on Television. Sage, 1995.

Madigan, Sheri, PhD; Dillon Browne, PhD; Nicole Racine, PhD. "Association Between Screen Time and Children's Performance on a Developmental Screening Test" *American Medical Association: JAMA Pediatr.* 2019;173(3):244-250. doi:10.1001/jamapediatrics.2018.5056. https://jamanetwork.com/journals/jamapediatrics/fullarticle/2722666, January 28, 2019.

Maker, Azmaira H. "Screen Time: The Impact on Kids and Parenting: New research explains the significant negative effects of excessive screen time." *Psychology Today*. https://www.psychologytoday.com/us/blog/helping-kids-cope/201808/screen-time-the-impact-kids-and-parenting, August 19, 2018.

Mangen, Anne; Bente R. Walgermo; Kolbjørn Brønnick. "Reading linear texts on paper versus computer screen: Effects on reading comprehension" *International Journal of Educational Research*, Volume 58, 2013, Pages 61–68. https://doi.org/10.1016/j.ijer.2012.12.002, January 5, 2013.

Marshall, Ron. "How Many Ads Do You See in One Day?" *Red Crow Marketing.* https://www.redcrowmarketing.com/2015/09/10/many-ads-see-one-day/, September 10, 2015.

"Masculine and Feminine Roles in Relationships (Sociology Essay)" UKEssays. http://www.ukessays.com/essays/sociology/masculine-and-feminine-roles-in-relationships-sociology-essay.php, November 2013, updates May 2017, November 2018.

Mathews, Jay. "Some kids need harder lessons than schools are willing to give them: New data reveals stubborn preference for below grade-level instruction" *The Washington Post.* https://www.washingtonpost.com/education/2022/08/21/grade-level-reading-difficulty-challenge/, August 21, 2022.

McCarthy, Caroline. "Hulu: We're evil, and proud of it" *CNet.* https://www.cnet.com/news/hulu-were-evil-and-proud-of-it/, February 2, 2009.

Michel, Dan, *Remorse of Conscience*, or *Ayenbite of inwyt*, ed. Richard Morris (London: N. Trubner & Co., 1895) pp. 74-75. Morris' 1895 republication is from Michel's 1340 translation from French to Kentish. Michel's 1340 translation is of the 13th century French *Somme le Roi.* My quote is from a translation of the 1340 Kentish into modern English, translation by Judith G. Humphries, in "The Personification of Death in Middle English Literature" (Denton, Texas: 1970, North Texas State University) p. 10.

Midling, Anne Sliper. Norwegian University of Science and Technology. "Why writing by hand makes kids smarter: Writing by hand creates much more activity in the sensorimotor parts of the brain, researchers found." *ScienceDaily.* https://www.sciencedaily.com/releases/2020/10/201001113540.htm, October 1, 2020.

Mischel, W., and E. B. Ebbesen (1970). "Attention in delay of gratification" *Journal of Personality and Social Psychology*, 16(2), 329–337. https://doi.org/10.1037/h0029815.

Montag, C., Lachmann, B., Herrlich, M., & Zweig, K. "Addictive Features of Social Media/Messenger Platforms and Freemium Games against the Background of Psychological and Economic Theories" *International journal of environmental research and public health*, *16*(14), 2612 (2019). https://doi.org/10.3390/ijerph16142612, July 23, 2019.

Mooney, Chris. "Liberals deny science, too" *The Washington Post.* https://www.washingtonpost.com/news/wonk/wp/2014/10/28/liberals-deny-science-too/, October 28, 2014.

Morrow, Jon. "How to Be Smart in a World of Dumb Bloggers" *BoostBlogTraffic*.
http://boostblogtraffic.com/smart-blogger/, September 17, 2013.
https://smartblogger.com/how-to-start-a-blog/, updated February 28, 2022.

Morrow, Jon. "On Gluttony, Selfishness, and Unleashing the Power Within" *BoostBlogTraffic*.
http://boostblogtraffic.com/unleash-your-power/, November 27, 2014.

"Most teenagers know smartphones compromise their lives, wish to curb use" *The Economic Times.*
https://economictimes.indiatimes.com/magazines/panache/most-teenagers-know-smartphones-compromise-their-lives-wish-to-curb-use/articleshow/64707105.cms, June 23, 2018.
Also see: Annie Sneed. "Are you addicted to your phone? Here is how to cut back" *The Economic Times*.
https://economictimes.indiatimes.com/news/how-to/are-you-addicted-to-your-phone-here-is-how-you-can-cut-back/articleshow/89607023.cms, February 16, 2022.

Nagata, Jason M. MD, MSc; Puja Iyer BA; Jonathan Chu BA; Fiona C. Baker PhD; Kelley Pettee Gabriel MS, PhD; Andrea K. Garber PhD, RD; Stuart B. Murray DClinPsych, PhD; Kirsten Bibbins-Domingo PhD, MD, MAS; Kyle T. Ganson PhD, MSW. "Contemporary screen time modalities among children 9–10 years old and binge-eating disorder at one-year follow-up: A prospective cohort study" *International Journal of Eating Disorders*, Vol. 54, Issue 5, pages 887–892. https://onlinelibrary.wiley.com/doi/10.1002/eat.23489, March 1, 2021.

Nagata, Jason M.; Puja Iyer; Jonathan Chu; Fiona C. Baker; Kelley Pettee Gabriel; Andrea K. Garber; Stuart B. Murray; Kirsten Bibbins-Domingo; Kyle T. Ganson. "Contemporary screen time usage among children 9–10-years-old is associated with higher body mass index percentile at 1-year follow-up: A prospective cohort study" *Pediatric Obesity*, Vol. 16, Issue 12. https://onlinelibrary.wiley.com/doi/10.1111/ijpo.12827, June 28, 2021.

Newman, Virginia. *Digging into Food Waste.* "Students from the Wenatchee School District take a closer look at the food served in their cafeteria, from the farm to the kitchen to the trays to the trash cans." https://youtu.be/91vU6DLHMPs, Published October 20, 2015.

Nichols, Tom. *The Death of Expertise.* New York: Oxford University Press, 2017.

Nielsen. "Percentage of Americans who say they watch too much TV: 49%" *BLS American Time Use Survey*, A.C. Nielsen Co. http://www.statisticbrain.com/television-watching-statistics/, Date Verified: 12.7.2013 (July 12, 2013).

Nielsen. "Television, Internet and Mobile Usage in the U.S. — A2/M2 Three Screen Report 4th Quarter 2008". Copyright © 2009 The Nielsen Company. http://i.cdn.turner.com/cnn/2009/images/02/24/screen.press.b.pdf, accessed May 29, 2014.

Owenz, Meghan. "The Rich Get Smart, The Poor Get Technology: The New Digital Divide in School Choice"*Screenfreeparenting.com*. https://www.screenfreeparenting.com/rich-get-smart-poor-get-technology-new-digital-divide-school-choice/, November 21, 2017.

Panksepp, Jaak. "Can PLAY diminish ADHD and facilitate the construction of the social brain?" *Journal of the Canadian Academy of Child and Adolescent Psychiatry* (16, 57-66), 2007.

Park, Alice. "Baby Einsteins: Not So Smart After All" *Time*. http://content.time.com/time/health/article/0,8599,1650352,00.html, August 06, 2007.

Park, Michael; Erin Leahey; Russell J. Funk. "Papers and patents are becoming less disruptive over time" *Nature* 613, 138–144 (2023). https://doi.org/10.1038/s41586-022-05543-x (https://www.nature.com/articles/s41586-022-05543-x), January 4, 2023.

Paul, Joe. "What is Nomophobia?" *TimeToLogOff.com*. https://www.itstimetologoff.com/2018/06/18/what-is-nomophobia/, June 2018.

Pellis, Sergio M., and Vivien C. Pellis. "Rough-and-Tumble Play: Training and Using the Social Brain" *The Oxford Handbook of the Development of Play* (December 2010).
Online Reference:
http://www.oxfordhandbooks.com/view/10.1093/oxfordhb/9780195393002.001.0001/oxfordhb-9780195393002-e-019, September 2012.

Pells, Rachael. "Today's four-year-olds often 'not physically ready' for school, experts warn" *Independent*. https://www.independent.co.uk/news/education/education-news/school-age-four-year-old-children-not-physically-ready-experts-warn-a7220476.html, September 1, 2016.

"Physical Fighting by Youth" *Child Trends Databank*. https://www.childtrends.org/indicators/physical-fighting-by-youth/, Accessed December 23, 2017.

Pickersgill, Eric. *Removed*. http://www.removed.social/, accessed October 21, 2015.

Place, Nathan. "Son accused of murdering parents after pretending to work at SpaceX" *The Independent*. https://www.independent.co.uk/news/world/americas/crime/son-murder-parents-spacex-wisconsin-b1988292.html, January 7, 2022.

"Planet or Plastic? A Brief History of How Plastic Has Changed Our World" *National Geographic.* https://video.nationalgeographic.com/video/magazine/planet-or-plastic/180516-ngm-brief-history-of-plastic, Accessed May 18, 2018.

Quast, Lisa. "Why Grit Is More Important Than IQ When You're Trying To Become Successful" *Forbes.* https://www.forbes.com/sites/lisaquast/2017/03/06/why-grit-is-more-important-than-iq-when-youre-trying-to-become-successful/, March 6, 2017.

Rampell, Catherine. "The rise of the 'gentleman's A' and the GPA arms race" *The Washington Post.* https://www.washingtonpost.com/opinions/the-rise-of-the-gentlemans-a-and-the-gpa-arms-race/2016/03/28/05c9e966-f522-11e5-9804-537defcc3cf6_story.html, March 28, 2016.

Randazzo, Sarah. "Schools Ditch Homework, Deadlines" *The Wall Street Journal*, Vol. CCLXXXI No. 97. Thursday April 27, 2023.

Rapaport, Lisa. "Screen time linked to ADHD symptoms in teens" *Reuters.* https://www.reuters.com/article/us-health-adhd-digital-media/screen-time-linked-to-adhd-symptoms-in-teens-idUSKBN1K72L8, July 17, 2018.

Redden, Elizabeth. "Foreign Student Dependence" *Inside Higher Education.* https://www.insidehighered.com/news/2013/07/12/new-report-shows-dependence-us-graduate-programs-foreign-students, July 12, 2013.

Reichel, Chloe. "The health effects of screen time on children: A research roundup" *Journalist's Resource.* https://journalistsresource.org/studies/society/public-health/screen-time-children-health-research/, May 14, 2019.

Richards, Rosalina; Rob McGee; Sheila M. Williams; David Welch; and Robert J. Hancox. "Adolescent Screen Time and Attachment to Parents and Peers" *Archives of Pediatrics & Adolescent Medicine.* 164, no. 3 (March 2010): 258–62. doi:10.1001/archpediatrics.2009.280. https://jamanetwork.com/journals/jamapediatrics/fullarticle/382905, March 1, 2010.

Richtel, Matt. "A Silicon Valley School That Doesn't Compute" *The New York Times.* https://www.nytimes.com/2011/10/23/technology/at-waldorf-school-in-silicon-valley-technology-can-wait.html, October 22, 2011.

Richtel, Matt. “The Myth of Multitasking” Attached to Technology and Paying a Price, p. 2.
Updated Title: “Your Brain on Computers: Attached to Technology and Paying a Price”
https://www.nytimes.com/2010/06/07/technology/07brain.html, June 6, 2010.

“The Rise of American Consumerism” *PBS.ORG.* https://www.pbs.org/wgbh/americanexperience/features/tupperware-consumer/, accessed December 15, 2018.

Rothman, Johanna. “Why Multitasking Doesn’t Work” *Pragmatic Manager*. https://www.jrothman.com/newsletter/2011/01/why-multitasking-doesnt-work/, January 1, 2011.

Ruiz, Rebecca R. “Betting Apps Can Make Anyone a Sports Fan. Even Me” *The New York Times*. https://www.nytimes.com/2022/12/24/business/sports-betting-apps.html, December 24, 2022.

Ryan, Claudine. “Why is watching TV so bad for you?” *ABC Health & Wellbeing*. http://www.abc.net.au/health/thepulse/stories/2014/07/24/4043618.htm, July 24, 2014.

Satariano, Adam. “British Ruling Pins Blame on Social Media for Teenager’s Suicide” *The New York Times.* https://www.nytimes.com/2022/10/01/business/instagram-suicide-ruling-britain.html, October 1, 2022.

Schor, Edward L. *Caring for your school-age child : ages 5 to 12.* American Academy of Pediatrics. New York : Bantam Books, 2004.
“What Children are NOT Doing When Watching TV” American Academy of Pediatrics.
http://www.healthychildren.org/English/family-life/Media/Pages/What-Children-are-NOT-Doing-When-Watching-TV.aspx, May 11, 2013.
Also See: *Caring for Our Children*. A Joint Collaborative Project of the American Academy of Pediatrics and the American Public Health Association.
https://nrckids.org/files/CFOC4%20pdf-%20FINAL.pdf, 2019.

Schulte, Brigid. *Overwhelmed: Work, Love, and Play When No One Has the Time.* New York: Simon & Schuster, Inc./Sarah Crichton Books, 2014.

Schuman, Rebecca. "Welcome to 13th Grade!" *Slate.com*. https://slate.com/human-interest/2014/10/high-schools-offer-a-fifth-year-13th-grade-is-a-great-idea.html, October 22, 2014.

Schwartz, Mel. "Is Our Society Manufacturing Depressed People?" *Psychology Today*. https://www.psychologytoday.com/blog/shift-mind/201203/is-our-society-manufacturing-depressed-people, March 19, 2012.

Sherman, L. E., Payton, A. A., Hernandez, L. M., Greenfield, P. M., & Dapretto, M. "The Power of the Like in Adolescence: Effects of Peer Influence on Neural and Behavioral Responses to Social Media" *Psychological science*, *27*(7), 1027–1035. https://doi.org/10.1177/0956797616645673, May 31, 2016.

Sigman, Aric. "How TV Is (Quite Literally) Killing Us" *Daily Mail*. http://www.whale.to/b/sigman.html, Oct 1, 2005.

Sikora, Joanna; M.D.R. Evans; Jonathan Kelley. "Scholarly culture: How books in adolescence enhance adult literacy, numeracy and technology skills in 31 societies" *Social Science Research*, January 2019, Volume 77, Pages 1–15. https://www.sciencedirect.com/science/article/abs/pii/S0049089X18300607 (https://doi.org/10.1016/j.ssresearch.2018.10.003), Available online October 2, 2018, Version of Record November 19, 2018.

Smith, Peter K. "Chapter 6: Physical Activity Play: Exercise Play and Rough-and-Tumble" *Children and Play* (April 2009).
Online Reference:
https://onlinelibrary.wiley.com/doi/10.1002/9781444311006.ch6, April 17, 2009.

Sornson, Bob. "Who's Looking Out for the Children?" *Early Learning Foundation*. http://earlylearningfoundation.com/whos-looking-out-for-the-children/, accessed March 18, 2017.

Sparks, Sarah D. “Students Are Behaving Badly in Class. Excessive Screen Time Might Be to Blame” Education Week. https://www.edweek.org/leadership/students-are-behaving-badly-in-class-excessive-screen-time-might-be-to-blame/2022/04, April 12, 2022.

Sriwilai, K., & Charoensukmongkol, P. “Face it, don’t Facebook it: impacts of social media addiction on mindfulness, coping strategies and the consequence on emotional exhaustion” *Stress and Health*, *32*(4), 427-434. https://doi.org/10.1002/smi.2637, March 30, 2015.

Steinmetz, Katy. “This Is What ‘Adulting’ Means” *Time Magazine.* https://time.com/4361866/adulting-definition-meaning/, June 8, 2016.

Stiglic, Neza and Russell M Viner. “Effects of screentime on the health and well-being of children and adolescents” *BMJ Journals: Pediatrics Research, Volume 9, Issue 1.* https://bmjopen.bmj.com/content/9/1/e023191, January 2019.

Stillman, Jessica. “Complaining Is Terrible for You, According to Science” *Inc.com*. http://www.inc.com/jessica-stillman/complaining-rewires-your-brain-for-negativity-science-says.html, February 29, 2016.

Stillman, Jessica. “The Cold, Hard Truth: You're Overwhelmed Because You Want to Be” *Inc.com.* http://www.inc.com/jessica-stillman/cure-for-feeling-overwhelmed-is-to-stop-talking-about-it.html, March 28, 2014.

Stillman, Jessica. “Which Country Has the Most Productive Workers?” *Inc.com.* http://www.inc.com/jessica-stillman/which-country-has-the-most-productive-workers.html, July 25, 2014.

Strasburger, Victor C., MD, FAAP, and Marjorie J. Hogan, MD, FAAP. “Children, Adolescents, and the Media”. American Academy of Pediatrics. http://pediatrics.aappublications.org/content/132/5/958.full, May 11, 2013.

Stromberg, Joseph. "Why you should take notes by hand — not on a laptop". *Vox: Science & Health*. http://www.vox.com/2014/6/4/5776804/note-taking-by-hand-versus-laptop, March 31, 2015.

Suckling, James and Jacquetta Lee. "Redefining scope: the true environmental impact of smartphones?" *The International Journal of Life Cycle Assessment*, Volume 20, Issue 8, pp 1181–1196. Springer Berlin Heidelberg. https://doi.org/10.1016/j.jclepro.2017.12.239, June 10, 2015.
Also Updated: Belkhir, Lotfi, and Ahmed Elmeligi. "Assessing ICT global emissions footprint: Trends to 2040 & recommendations" *Journal of Cleaner Production*. Volume 177, pp 448-463. https://www.sciencedirect.com/science/article/abs/pii/S095965261733233X, Online January 2, 2018; Version of Record January 3, 2018.

Szalavitz, Maia. "Misery Has More Company Than You Think, Especially on Facebook". *Time*. http://healthland.time.com/2011/01/27/youre-not-alone-misery-has-more-company-than-you-think/, January 27, 2011.

Takeuchi, Hikaru; Yasuyuki Taki; Hiroshi Hashizume; Kohei Asano; Michiko Asano; Yuko Sassa; Susumu Yokota; Yuka Kotozaki; Rui Nouchi; and Ryuta Kawashima. "The Impact of Television Viewing on Brain Structures: Cross-Sectional and Longitudinal Analyses" *Cerebral Cortex*, Volume 25, Issue 5, Pages 1188–1197, May 2015, Oxford Journals. https://cercor.oxfordjournals.org/content/early/2013/11/18/cercor.bht315.full, April 15, 2015.
https://academic.oup.com/cercor/article/25/5/1188/311796, May 2015.
https://doi.org/10.1093/cercor/bht315, May 2015.

Tamana, Sukhpreet K.; Victor Ezeugwu; Joyce Chikuma; Diana L. Lefebvre; Meghan B. Azad; Theo J. Moraes; Padmaja Subbarao; Allan B. Becker; Stuart E. Turvey; Malcolm R. Sears; Bruce D. Dick; Valerie Carson; Carmen Rasmussen; Piush J. Mandhane. "Screen-time is associated with inattention problems in preschoolers: Results from the CHILD birth cohort study" *PLOS ONE Journal of Scientific Research*. https://journals.plos.org/plosone/article?id=10.1371/journal.pone.0213995, April 17, 2019.

Taylor, Jim. "Parenting: The Sad Misuse of Self-esteem" *Psychology Today.* https://www.psychologytoday.com/blog/the-power-prime/201002/parenting-the-sad-misuse-self-esteem, February 22, 2010.

Taylor, Jim. "Pro Athletes are Hooked on Social Media Too" https://www.drjimtaylor.com/4.0/pro-athletes-are-hooked-on-social-media-too/, March 26, 2018.

"Television Addiction" *All About Life Challenges*. http://www.allaboutlifechallenges.org/television-addiction.htm, accessed September 21, 2014.

"Television History - The First 75 Years" *TVhistory.TV.* © 2001-2013. http://www.tvhistory.tv/facts-stats.htm.

"Television: Opiate of the Masses". *FamilyResource.com*. http://www.familyresource.com/lifestyles/mental-environment/television-opiate-of-the-masses, accessed May 29, 2014.

"TV retards your child's development" *Consumers Association of Penang*. http://www.consumer.org.my/index.php/development/education/347-tv-retards-your-childs-development, accessed July 17, 2016.
Also See:
"More arguments against TV" Consumers Association of Penang. https://consumer.org.my/more-arguments-against-tv/, March 18, 2022.

"Television vs. Reading" Parent Soup®, a Trademark of iVillagesm Inc. Copyright 1996.
http://webshare.northseattle.edu/fam180/topics/tv/tvvsread.htm, accessed May 29, 2014.

Thompson, Alexandra. "Surgery students are struggling to use their hands because they spent 'too much time watching TV' rather than sewing or playing an instrument as children" *The Daily Mail.* https://www.dailymail.co.uk/health/article-6332847/Surgery-students-struggle-use-hands-spent-time-watching-TV.html, October 30, 2018.

Thompson, Dennis. "More Americans suffering from stress, anxiety and depression, study finds" *CBS News - Healthday.* http://www.cbsnews.com/news/stress-anxiety-depression-mental-illness-increases-study-finds/, April 17, 2017.

Tintocalis, Ana. "San Francisco Middle Schools No Longer Teaching 'Algebra 1'" *The California Report*, KQED News. https://www.kqed.org/news/10610214/san-francisco-middle-schools-no-longer-teaching-algebra-1, July 22, 2015.

TNTP. "Accelerate, Don't Remediate: New Evidence from Elementary Math Classrooms" *The New Teacher Project* (tntp.org). https://tntp.org/publications/view/teacher-training-and-classroom-practice/accelerate-dont-remediate, https://tntp.org/assets/documents/TNTP_Accelerate_Dont_Remediate_FINAL.pdf, May 23, 2021.

TNTP. "Unlocking Acceleration: How Below Grade-Level Work is Holding Students Back in Literacy" *The New Teacher Project* (tntp.org). https://tntp.org/assets/documents/Unlocking_Acceleration_8.16.22.pdf, August 2022.

"Top 10 Best Clark County Public Schools (2023)" *Nevada Public School Review*. https://www.publicschoolreview.com/nevada/clark-county, accessed April 28, 2023.

Toppo, Greg. "Americans trail adults in other countries in math, literacy, problem-solving" *USA Today*. http://www.usatoday.com/story/news/nation/2013/10/08/literacy-international-workers-education-math-americans/2935909/, October 8, 2013.

Trelease, Jim. *The Read-Aloud Handbook*, Penguin Books, 1995.

Turner, Cory. "The Surgeon Who Became An Activist For Baby Talk" *NPR Ed.* https://www.npr.org/sections/ed/2015/09/14/437515492/the-surgeon-who-became-an-activist-for-baby-talk, September 14, 2015.

Twenge, Jean M. “Changes in how we’re spending our free time might explain the unhappiness epidemic” *PsyPost.org*. https://www.psypost.org/2018/08/changes-in-how-were-spending-our-free-time-might-explain-the-unhappiness-epidemic-51893, August 5, 2018.

Twenge, Jean M. “Have Smartphones Destroyed a Generation?” *The Atlantic*. https://www.theatlantic.com/magazine/archive/2017/09/has-the-smartphone-destroyed-a-generation/534198/, September 2017.

Twenge, Jean M. “This Is What Happy Teens Do. Hint: It doesn’t involve their phones.” *Psychology Today*. https://www.psychologytoday.com/us/blog/our-changing-culture/201808/is-what-happy-teens-do, August 31, 2018.

Twenge, Jean M.; G. N. Martin, W. K. Campbell. “Decreases in psychological well-being among American adolescents after 2012 and links to screen time during the rise of smartphone technology.” American Psychological Association Journal Emotion, 18(6), 765–780. https://doi.org/10.1037/emo0000403, 2018.

Twenge, Jean M.; Thomas E. Joiner; Megan L. Rogers; Gabrielle N. Martin. “Increases in Depressive Symptoms, Suicide-Related Outcomes, and Suicide Rates Among U.S. Adolescents After 2010 and Links to Increased New Media Screen Time” *Clinical Psychological Science*. http://journals.sagepub.com/doi/full/10.1177/2167702617723376, November 14, 2017.

Twenge, Jean M. and W. Keith Campbell. “Associations between screen time and lower psychological well-being among children and adolescents: Evidence from a population-based study” *Preventive Medicine Reports,* Volume 12, Pages 271-283. Elsevier B.V. https://doi.org/10.1016/j.pmedr.2018.10.003, October 18, 2018.

University of Tsukuba. “Active body, active mind: The secret to a younger brain may lie in exercising your body” *ScienceDaily*. https://www.sciencedaily.com/releases/2015/10/151023084456.htm, October 23, 2015.

van den Heuvel, Meta; Julia Ma; Cornelia M Borkhoff; Christine Koroshegyi; David W H Dai; Patricia C Parkin; Jonathon L Maguire; Catherine S Birken. “Mobile Media Device Use is Associated with Expressive Language Delay in 18-Month-Old Children.” *Journal of developmental and behavioral pediatrics : JDBP* vol. 40,2 (2019): 99-104. doi:10.1097/DBP.0000000000000630. https://pubmed.ncbi.nlm.nih.gov/30753173/, Feb/Mar 2019.

van der Weel, F. R. (Ruud); Audrey L. H. van der Meer. “Handwriting but not typewriting leads to widespread brain connectivity: a high-density EEG study with implications for the classroom.” *Frontiers in Psychology, Sec. Educational Psychology*, Volume 14 - 2023.
https://doi.org/10.3389/fpsyg.2023.1219945
https://www.frontiersin.org/journals/psychology/articles/10.3389/fpsyg.2023.1219945/full, January 25, 2024.

Vedantam, Shankar. “There’s A Gap Between Perception And Reality When It Comes To Learning” *NPR Science*.
https://www.npr.org/2019/02/18/695637906/theres-a-gap-between-perception-and-reality-when-it-comes-to-learning, February 18, 2019.

Walter, Tatumn. “A tenth of each day spent watching television” *USC US-China Institute*.
https://china.usc.edu/tenth-each-day-spent-watching-television, accessed May 29, 2014.

Wang, Pengcheng; Jia Nie; Xingchao Wang; Yuhui Wang; Fengqing Zhao; Xiaochun Xie; Li Lei; Mingkun Ouyang,. “How are smartphones associated with adolescent materialism?” *Journal of Health Psychology* Vol. 25 (September 19, 2018).
https://journals.sagepub.com/doi/10.1177/1359105318801069.

“Want a younger brain? Stay in school — and take the stairs” *Science Daily*.
https://www.sciencedaily.com/releases/2016/03/160309125520.htm March 9, 2016.

Ward, Adrian F.; Kristen Duke; Ayelet Gneezy; and Maarten W. Bos. "Brain Drain: The Mere Presence of One's Own Smartphone Reduces Available Cognitive Capacity". *Journal of the Association for Consumer Research* 2, no. 2 (April 2017): 140-154.
https://doi.org/10.1086/691462, April 3, 2017.

Warner, Jeremy. "Harsh truths about the decline of Britain" T*he Telegraph.* http://www.telegraph.co.uk/finance/economics/10417838/Harsh-truths-about-the-decline-of-Britain.html, October 31, 2013.

Weller, Chris. "An MIT psychologist explains why so many tech moguls send their kids to anti-tech schools" *Business Insider.*
https://www.businessinsider.com/sherry-turkle-why-tech-moguls-send-their-kids-to-anti-tech-schools-2017-11, November 7, 2017.

Williams, Ray. "Anti-Intellectualism and the 'Dumbing Down' of America" Indiana University Center for Civic Literacy.
https://blogs.iu.edu/civicliteracy/2015/05/22/anti-intellectualism-and-the-dumbing-down-of-america/, July 7, 2014, Online May 22, 2015.

Wilson, Hugh. "Are men becoming more feminine?" *MSN.com*.
http://him.uk.msn.com/in-the-know/are-men-becoming-more-feminine-women-male-female-gender-gap, July 19, 2013.

Winn, Marie. *Plug-In Drug* (New York: Viking Penguin, 1977, Revised and Updated Edition, 2002).

Wolf, Maryanne. "Skim reading is the new normal. The effect on society is profound" *The Guardian*.
https://www.theguardian.com/commentisfree/2018/aug/25/skim-reading-new-normal-maryanne-wolf, August 25, 2018.

Woods, H. C., & Scott, H. "# Sleepyteens: Social media use in adolescence is associated with poor sleep quality, anxiety, depression and low self-esteem" *Journal of adolescence*, *51*, 41-49.
https://doi.org/10.1016/j.adolescence.2016.05.008, June 10, 2016.

Woody, Maxwell et al. "Corrigendum: The role of pickup truck electrification in the decarbonization of light-duty vehicles" *Environmental Research Letters* 17 089501. *IOP Publishing Ltd.* https://iopscience.iop.org/article/10.1088/1748-9326/ac7cfc, July 15, 2022.

Younas, Adeel. "Report: Smartphones Are Destroying Our Planet Faster Than We Think" *TechWafer.* https://techwafer.com/report-smartphones-are-destroying-our-planet-faster-than-we-think/, April 2, 2018.

Yuan, Kai; Ping Cheng; Tao Dong; et al. "Cortical Thickness Abnormalities in Late Adolescence with Online Gaming Addiction" *PLOS ONE.* https://doi.org/10.1371/journal.pone.0053055, January 9, 2013.

Yuan, Kai; Wei Qin; Guihong Wang; et al. "Microstructure Abnormalities in Adolescents with Internet Addiction Disorder" *PLOS ONE.* https://doi.org/10.1371/journal.pone.0020708, June 3, 2011.

Zaki, Jamil. "What, Me Care? Young Are Less Empathetic" A recent study finds a decline in empathy among young people in the U.S. *Scientific American.* http://www.scientificamerican.com/article/what-me-care/, December 23, 2010.

END NOTES

1

- Herbert E. Krugman and Eugene L. Hartley. "Passive Learning From Television" *The Public Opinion Quarterly,* Vol. 34, No. 2. (Oxford University Press, 1970) pp. 184-190.
- Herbert E. Krugman. "Brain Wave Measures of Media Involvement" *How Advertising Works: The Role of Research.* (New York SAGE Publications, 1998) pp. 139–151.

2

- Herbert E. Krugman and Eugene L. Hartley. "Passive Learning From Television" *The Public Opinion Quarterly,* Vol. 34, No. 2. (Oxford University Press, 1970) pp. 184-190.
- Herbert E. Krugman. "Brain Wave Measures of Media Involvement" *How Advertising Works: The Role of Research.* (New York SAGE Publications, 1998) pp. 139–151.
- "Your brain waves change when you watch TV" *I Am Awake.* http://www.iamawake.co/your-brain-waves-change-when-you-watch-tv/, October 11, 2013.
- Allan Stromfeldt Christensen. "Lemminged: to be herded off the peak oil cliff by filmmakers" *TransitionVoice.com.* http://transitionvoice.com/2014/11/lemminged-to-be-herded-off-the-peak-oil-cliff-by-filmmakers/, November 11, 2014.
- "Television: Opiate of the Masses" *FamilyResource.com*. http://www.familyresource.com/lifestyles/mental-environment/television-opiate-of-the-masses, accessed May 29, 2014.
- Douglas Fields. "Watching TV Alters Children's Brain Structure and Lowers IQ". http://rdouglasfields.com/2015/05/watching-tv-alters-childrens-brain-structure-and-lowers-iq/, May 4, 2015.
- Kristina Birdsong. "This is Your Child's Brain on TV" *Scientific Learning.* http://www.scilearn.com/blog/how-television-impacts-learning, March 22, 2016. http://54.186.226.228/blog/your-childs-brain-tv, March 22, 2016.

3

- “TV retards your child’s development” *Consumers Association of Penang.* http://www.consumer.org.my/index.php/development/education/347-tv-retards-your-childs-development, accessed July 17, 2016.
  Also See:
  “More arguments against TV” Consumers Association of Penang. https://consumer.org.my/more-arguments-against-tv/, March 18, 2022.
- Emma Henderson. “Watching lots of TV ‘makes you stupid’” *The Independent.* http://www.independent.co.uk/news/science/watching-lots-of-tv-makes-you-stupid-says-american-universities-a6759026.html, December 3, 2015.
- Herbert E. Krugman and Eugene L. Hartley. “Passive Learning From Television” *The Public Opinion Quarterly,* Vol. 34, No. 2. (Oxford University Press, 1970) pp. 184-190.
- Herbert E. Krugman. “Brain Wave Measures of Media Involvement” *How Advertising Works: The Role of Research.* (New York SAGE Publications, 1998) pp. 139–151.
- “Your brain waves change when you watch TV” *I Am Awake.* http://www.iamawake.co/your-brain-waves-change-when-you-watch-tv/, October 11, 2013.
- Kristina Birdsong. “This is Your Child’s Brain on TV” *Scientific Learning.* http://www.scilearn.com/blog/how-television-impacts-learning, March 22, 2016.
  http://54.186.226.228/blog/your-childs-brain-tv, March 22, 2016.
- Douglas Fields. “Watching TV Alters Children’s Brain Structure and Lowers IQ”. http://rdouglasfields.com/2015/05/watching-tv-alters-childrens-brain-structure-and-lowers-iq/, May 4, 2015.
- Jamil Zaki. “What, Me Care? Young Are Less Empathetic” *Scientific American*. A recent study finds a decline in empathy among young people in the U.S. http://www.scientificamerican.com/article/what-me-care/, December 23, 2010.

- "Children who watch 'excessive' amounts of TV are more likely to have criminal convictions, exhibit aggression and experience negative emotions: study." *New York Daily News*. http://www.nydailynews.com/life-style/health/kids-watch-excessive-tv-criminal-convictions-young-adulthood-study-article-1.1267868, February 19, 2013.
"With every hour in front of the television, kids were more likely to show aggressive behavior or receive a criminal conviction by early adulthood, according to a study published in 'Pediatrics.' The issue isn't necessarily the content of the programming, but the social isolation that comes from so many hours in front of the tube."
- David L. Hill. "Why to Avoid TV Before Age 2" (early brain development). *American Academy of Pediatrics*. http://www.healthychildren.org/English/family-life/Media/Pages/Why-to-Avoid-TV-Before-Age-2.aspx, May 11, 2013.
Original Source: Dad to Dad: Parenting Like a Pro (Copyright © American Academy of Pediatrics 2012).
- Alice Park. "Baby Einsteins: Not So Smart After All" *Time*. http://content.time.com/time/health/article/0,8599,1650352,00.html August 06, 2007.
- "Television vs. Reading" Parent Soup®, a Trademark of iVillagesm Inc. Copyright 1996. http://webshare.northseattle.edu/fam180/topics/tv/tvvsread.htm, accessed May 29, 2014.
- Jim Trelease. *The Read-Aloud Handbook*, Penguin Books, 1995.
- Marie Winn. "Television and the Brain" and "Brain Changes" *Plug-In Drug.* (New York: Viking Penguin, 1977, Revised and Updated Edition, 2002) pp. 67–69:
"'Research conducted during the next two decades removed any doubt about the impact of early brain stimulation on a child's later cognitive development. ... they were able to demonstrate that environmental factors can alter neuron pathways during early childhood and long after ... among the most important of the environmental factors ... are the language and eye contact an infant is exposed to [and] ... the number of words an infant hears each day is the single most important predictor of later intelligence, school success and social competence. But there's one catch. The words have to come from an attentive, engaged human being ... radio and television do not work."

4

- David Hinckley. "Average American watches 5 hours of TV per day, report shows" *New York Daily News*. https://www.nydailynews.com/life-style/average-american-watches-5-hours-tv-day-article-1.1711954, March 5, 2014.
  Per Nielsen, here's the average weekly usage for ascending age groups. Ages:
  - 2–11: 24 hours, 16 minutes. (round down to 24 hours, 15 minutes, or 24.25 x 10 years counting ages 2 and 11 = 12,610)
  - 12–17: 20 hours, 41 minutes. (round down to 20 hours, 30 minutes, or 20.5 hours x 6 years counting ages 12 and 17 = 6,396)
  - 18–24: 22 hours, 27 minutes. (round down to 22 hours, 20 minutes, or 20.33 hours x 6 years counting age 18 to end of age 23 = 6,343)
  - By age 24, the average American has watched 25,349 hours of TV (does not include time on other devices). Here is the breakdown continuing older age groups, showing people watch more and more as they get older. Ages:
    - 25–34: 27 hours, 36 minutes.
    - 35–49: 33 hours, 40 minutes.
    - 50–64: 43 hours, 56 minutes.
    - 65–plus: 50 hours, 34 minutes.
- Victor C. Strasburger, MD, FAAP, and Marjorie J. Hogan, MD, FAAP. "Children, Adolescents, and the Media" *American Academy of Pediatrics*. http://pediatrics.aappublications.org/content/132/5/958.full, May 11, 2013.

5

- Tom Nichols. *The Death of Expertise.* p. 140. New York: Oxford University Press, 2017.

6

- Victor C. Strasburger, MD, FAAP, and Marjorie J. Hogan, MD, FAAP. "Children, Adolescents, and the Media". American Academy of Pediatrics. http://pediatrics.aappublications.org/content/132/5/958.full, May 11, 2013.

7

- Alice Park. "Baby Einsteins: Not So Smart After All" *Time*. http://content.time.com/time/health/article/0,8599,1650352,00.html August 06, 2007.
- "Television vs. Reading" *Parent Soup*®, a Trademark of iVillage℠ Inc. Copyright 1996. http://webshare.northseattle.edu/fam180/topics/tv/tvvsread.htm, accessed May 29, 2014.
- Jim Trelease. *The Read-Aloud Handbook*, Penguin Books, 1995.

8

- Robert Kubey and Mihaly Csikszentmihalyi. "Television Addiction is no mere metaphor" *Scientific American.* http://www.academia.edu/5065840/Television_Addiction_is_no_mere_metaphor, accessed January 23, 2016.

9

- Herbert E. Krugman and Eugene L. Hartley. "Passive Learning From Television" *The Public Opinion Quarterly,* Vol. 34, No. 2. (Oxford University Press, 1970) pp. 184-190.
- Herbert E. Krugman. "Brain Wave Measures of Media Involvement" *How Advertising Works: The Role of Research.* (New York SAGE Publications, 1998) pp. 139–151.
- "Your brain waves change when you watch TV" *I Am Awake.* http://www.iamawake.co/your-brain-waves-change-when-you-watch-tv/, October 11, 2013.
- Allan Stromfeldt Christensen. "Lemminged: to be herded off the peak oil cliff by filmmakers" *TransitionVoice.com.* http://transitionvoice.com/2014/11/lemminged-to-be-herded-off-the-peak-oil-cliff-by-filmmakers/, November 11, 2014.
- "Television: Opiate of the Masses" *FamilyResource.com*. http://www.familyresource.com/lifestyles/mental-environment/television-opiate-of-the-masses, accessed May 29, 2014.
- Douglas Fields. "Watching TV Alters Children's Brain Structure and Lowers IQ". http://rdouglasfields.com/2015/05/watching-tv-alters-childrens-brain-structure-and-lowers-iq/, May 4, 2015.

* Kristina Birdsong. "This is Your Child's Brain on TV" *Scientific Learning.* http://www.scilearn.com/blog/how-television-impacts-learning, March 22, 2016.
  http://54.186.226.228/blog/your-childs-brain-tv, March 22, 2016.

10

* Shankar Vedantam. "There's A Gap Between Perception And Reality When It Comes To Learning" *NPR Science.* https://www.npr.org/2019/02/18/695637906/theres-a-gap-between-perception-and-reality-when-it-comes-to-learning, February 18, 2019.

11

* Shankar Vedantam. "There's A Gap Between Perception And Reality When It Comes To Learning" *NPR Science.* https://www.npr.org/2019/02/18/695637906/theres-a-gap-between-perception-and-reality-when-it-comes-to-learning, February 18, 2019.

12

* "TV retards your child's development" *Consumers Association of Penang.* http://www.consumer.org.my/index.php/development/education/347-tv-retards-your-childs-development, accessed July 17, 2016.
  Also See:
  "More arguments against TV" Consumers Association of Penang. https://consumer.org.my/more-arguments-against-tv/, March 18, 2022.
* Emma Henderson. "Watching lots of TV 'makes you stupid'" *The Independent.* http://www.independent.co.uk/news/science/watching-lots-of-tv-makes-you-stupid-says-american-universities-a6759026.html, December 3, 2015.
* Kristina Birdsong. "This is Your Child's Brain on TV" *Scientific Learning.* http://www.scilearn.com/blog/how-television-impacts-learning, March 22, 2016.
  http://54.186.226.228/blog/your-childs-brain-tv, March 22, 2016.
* Herbert E. Krugman and Eugene L. Hartley. "Passive Learning From Television" *The Public Opinion Quarterly,* Vol. 34, No. 2. (Oxford University Press, 1970) pp. 184-190.
* Herbert E. Krugman. "Brain Wave Measures of Media Involvement" *How Advertising Works: The Role of Research.* (New York SAGE Publications, 1998) pp. 139–151.

- "Your brain waves change when you watch TV" *I Am Awake.* http://www.iamawake.co/your-brain-waves-change-when-you-watch-tv/, October 11, 2013.
- Kristina Birdsong. "This is Your Child's Brain on TV" *Scientific Learning.* http://www.scilearn.com/blog/how-television-impacts-learning, March 22, 2016.
  http://54.186.226.228/blog/your-childs-brain-tv, March 22, 2016.
- Douglas Fields. "Watching TV Alters Children's Brain Structure and Lowers IQ".
  http://rdouglasfields.com/2015/05/watching-tv-alters-childrens-brain-structure-and-lowers-iq/, May 4, 2015.
- Jamil Zaki. "What, Me Care? Young Are Less Empathetic" *Scientific American*. A recent study finds a decline in empathy among young people in the U.S.
  http://www.scientificamerican.com/article/what-me-care/, December 23, 2010.
- "Children who watch 'excessive' amounts of TV are more likely to have criminal convictions, exhibit aggression and experience negative emotions: study." *New York Daily News*.
  http://www.nydailynews.com/life-style/health/kids-watch-excessive-tv-criminal-convictions-young-adulthood-study-article-1.1267868, February 19, 2013.
  "With every hour in front of the television, kids were more likely to show aggressive behavior or receive a criminal conviction by early adulthood, according to a study published in 'Pediatrics.' The issue isn't necessarily the content of the programming, but the social isolation that comes from so many hours in front of the tube."
  David L. Hill. "Why to Avoid TV Before Age 2" (early brain development). *American Academy of Pediatrics*.
  http://www.healthychildren.org/English/family-life/Media/Pages/Why-to-Avoid-TV-Before-Age-2.aspx, May 11, 2013.
  Original Source: Dad to Dad: Parenting Like a Pro (Copyright © American Academy of Pediatrics 2012).
- Alice Park. "Baby Einsteins: Not So Smart After All" *Time*.
  http://content.time.com/time/health/article/0,8599,1650352,00.html August 06, 2007.

- "Television vs. Reading" *Parent Soup*®, a Trademark of iVillage℠ Inc. Copyright 1996. http://webshare.northseattle.edu/fam180/topics/tv/tvvsread.htm, accessed May 29, 2014.
- Jim Trelease. *The Read-Aloud Handbook*, Penguin Books, 1995.
- Edward L. Schor. *Caring for your school-age child : ages 5 to 12.* American Academy of Pediatrics. New York : Bantam Books, 2004.
  "What Children are NOT Doing When Watching TV" American Academy of Pediatrics.
  http://www.healthychildren.org/English/family-life/Media/Pages/What-Children-are-NOT-Doing-When-Watching-TV.aspx, May 11, 2013.
  Also See: *Caring for Our Children*. A Joint Collaborative Project of the American Academy of Pediatrics and the American Public Health Association.
  https://nrckids.org/files/CFOC4%20pdf-%20FINAL.pdf, 2019.
- Tina D. Hoang, MSPH; Jared Reis, PhD; Na Zhu, MD, MPH; David R. Jacobs Jr, PhD; Lenore J. Launer, PhD; Rachel A. Whitmer, PhD; Stephen Sidney, MD; Kristine Yaffe, MD.
  "Effect of Early Adult Patterns of Physical Activity and Television Viewing on Midlife Cognitive Function" *JAMA Psychiatry* (The Journal of the American Medical Association).
  http://archpsyc.jamanetwork.com/article.aspx?articleid=2471270, January 2016, Vol 73, No. 1.
- Marie Winn. "Television and the Brain," "Brain Changes," "Losing the Thread," "The Basic Building Blocks," "A Preference for Watching," "Free Time and Resourcefulness" *Plug-In Drug*. (New York: Viking Penguin, 1977, Revised and Updated Edition, 2002) pp. 67–69, 95–99, 131.
- Kyung Hee Kim. "The Creativity Crisis: The Decrease in Creative Thinking Scores on the Torrance Tests of Creative Thinking" *Creativity Research Journal*, 23:4, 285–295. Routledge, Taylor & Francis Group. https://www.tandfonline.com/doi/abs/10.1080/10400419.2011.627805, November 9, 2011.
  Downloaded by the College of William and Mary: https://kkim.wmwikis.net/file/view/Kim_2011_Creativity_crisis.pdf, April 3, 2012.

13

- Herbert E. Krugman and Eugene L. Hartley. "Passive Learning From Television" *The Public Opinion Quarterly,* Vol. 34, No. 2. (Oxford University Press, 1970) pp. 184-190.
- Herbert E. Krugman. "Brain Wave Measures of Media Involvement" *How Advertising Works: The Role of Research.* (New York SAGE Publications, 1998) pp. 139–151.
- "Your brain waves change when you watch TV" *I Am Awake.* http://www.iamawake.co/your-brain-waves-change-when-you-watch-tv/, October 11, 2013.

14

- Tom Leonard. "'Passive' TV Can Harm Your Baby's Speech Making It Harder for Them to Later Cope In School" *Daily Mail.* http://www.dailymail.co.uk/news/article-2054950/Passive-TV-watching-harm-babies-speech.html, October 28, 2011.
- Daile Cross. "Treat TV like passive smoking when it comes to children" *WA Today.* https://www.watoday.com.au/national/western-australia/treat-tv-like-passive-smoking-when-it-comes-to-children-20171106-gzftv4.html, November 9, 2017.

15

- Kristina Birdsong. "This is Your Child's Brain on TV" *Scientific Learning.* http://www.scilearn.com/blog/how-television-impacts-learning, March 22, 2016.
  http://54.186.226.228/blog/your-childs-brain-tv, March 22, 2016.
- Tom Leonard. "'Passive' TV Can Harm Your Baby's Speech Making It Harder for Them to Later Cope In School" *Daily Mail.* http://www.dailymail.co.uk/news/article-2054950/Passive-TV-watching-harm-babies-speech.html, October 28, 2011.
- Daile Cross. "Treat TV like passive smoking when it comes to children" *WA Today.* https://www.watoday.com.au/national/western-australia/treat-tv-like-passive-smoking-when-it-comes-to-children-20171106-gzftv4.html, November 9, 2017.

16

- Joanna Sikora; M.D.R. Evans; Jonathan Kelley. "Scholarly culture: How books in adolescence enhance adult literacy, numeracy and technology skills in 31 societies" *Social Science Research*, January 2019, Volume 77, Pages 1–15. https://www.sciencedirect.com/science/article/abs/pii/S0049089X18300607 (https://doi.org/10.1016/j.ssresearch.2018.10.003), Available online October 2, 2018, Version of Record November 19, 2018.
- Robby Berman. "A home library can have a powerful effect on children: A new study finds that simply growing up in a home with enough books increases adult literacy and math prowess." *BigThink*. https://bigthink.com/the-present/mind-brain-home-library-benefits, October 13, 2018.

17

- Travis Bradberry and Jean Greaves. *Emotional Intelligence 2.0.* pp. 51, 52. San Diego, California: TalentSmart, 2009.

18

- "TV retards your child's development" *Consumers Association of Penang.* http://www.consumer.org.my/index.php/development/education/347-tv-retards-your-childs-development, accessed July 17, 2016.
  Also See:
  "More arguments against TV" Consumers Association of Penang. https://consumer.org.my/more-arguments-against-tv/, March 18, 2022.
- Emma Henderson. "Watching lots of TV 'makes you stupid'" *The Independent.* http://www.independent.co.uk/news/science/watching-lots-of-tv-makes-you-stupid-says-american-universities-a6759026.html, December 3, 2015.
- Kristina Birdsong. "This is Your Child's Brain on TV" *Scientific Learning.* http://www.scilearn.com/blog/how-television-impacts-learning, March 22, 2016.
  http://54.186.226.228/blog/your-childs-brain-tv, March 22, 2016.
- Herbert E. Krugman and Eugene L. Hartley. "Passive Learning From Television" *The Public Opinion Quarterly,* Vol. 34, No. 2. (Oxford University Press, 1970) pp. 184-190.
- Herbert E. Krugman. "Brain Wave Measures of Media Involvement" *How Advertising Works: The Role of Research.* (New York SAGE Publications, 1998) pp. 139–151.

- "Your brain waves change when you watch TV" *I Am Awake.* http://www.iamawake.co/your-brain-waves-change-when-you-watch-tv/, October 11, 2013.
- Kristina Birdsong. "This is Your Child's Brain on TV" *Scientific Learning.* http://www.scilearn.com/blog/how-television-impacts-learning, March 22, 2016.
  http://54.186.226.228/blog/your-childs-brain-tv, March 22, 2016.
- Douglas Fields. "Watching TV Alters Children's Brain Structure and Lowers IQ".
  http://rdouglasfields.com/2015/05/watching-tv-alters-childrens-brain-structure-and-lowers-iq/, May 4, 2015.
- Jamil Zaki. "What, Me Care? Young Are Less Empathetic" *Scientific American*. A recent study finds a decline in empathy among young people in the U.S.
  http://www.scientificamerican.com/article/what-me-care/, December 23, 2010.
- "Children who watch 'excessive' amounts of TV are more likely to have criminal convictions, exhibit aggression and experience negative emotions: study." *New York Daily News*.
  http://www.nydailynews.com/life-style/health/kids-watch-excessive-tv-criminal-convictions-young-adulthood-study-article-1.1267868, February 19, 2013.
  "With every hour in front of the television, kids were more likely to show aggressive behavior or receive a criminal conviction by early adulthood, according to a study published in 'Pediatrics.' The issue isn't necessarily the content of the programming, but the social isolation that comes from so many hours in front of the tube."
  David L. Hill. "Why to Avoid TV Before Age 2" (early brain development). *American Academy of Pediatrics*.
  http://www.healthychildren.org/English/family-life/Media/Pages/Why-to-Avoid-TV-Before-Age-2.aspx, May 11, 2013.
  Original Source: Dad to Dad: Parenting Like a Pro (Copyright © American Academy of Pediatrics 2012).
- Alice Park. "Baby Einsteins: Not So Smart After All" *Time*. http://content.time.com/time/health/article/0,8599,1650352,00.html August 06, 2007.

- "Television vs. Reading" *Parent Soup*®, a Trademark of iVillage℠ Inc. Copyright 1996.
  http://webshare.northseattle.edu/fam180/topics/tv/tvvsread.htm, accessed May 29, 2014.
- Jim Trelease. *The Read-Aloud Handbook*, Penguin Books, 1995.
- Edward L. Schor. *Caring for your school-age child : ages 5 to 12.* American Academy of Pediatrics. New York : Bantam Books, 2004.
  "What Children are NOT Doing When Watching TV" American Academy of Pediatrics.
  http://www.healthychildren.org/English/family-life/Media/Pages/What-Children-are-NOT-Doing-When-Watching-TV.aspx, May 11, 2013.
  Also See: *Caring for Our Children*. A Joint Collaborative Project of the American Academy of Pediatrics and the American Public Health Association.
  https://nrckids.org/files/CFOC4%20pdf-%20FINAL.pdf, 2019.
- Tina D. Hoang, MSPH; Jared Reis, PhD; Na Zhu, MD, MPH; David R. Jacobs Jr, PhD; Lenore J. Launer, PhD; Rachel A. Whitmer, PhD; Stephen Sidney, MD; Kristine Yaffe, MD.
  "Effect of Early Adult Patterns of Physical Activity and Television Viewing on Midlife Cognitive Function" *JAMA Psychiatry* (The Journal of the American Medical Association).
  http://archpsyc.jamanetwork.com/article.aspx?articleid=2471270, January 2016, Vol 73, No. 1.
- Marie Winn. "Television and the Brain," "Brain Changes," "Losing the Thread," "The Basic Building Blocks," "A Preference for Watching," "Free Time and Resourcefulness" *Plug-In Drug*. (New York: Viking Penguin, 1977, Revised and Updated Edition, 2002) pp. 67–69, 95–99, 131.
- Kyung Hee Kim. "The Creativity Crisis: The Decrease in Creative Thinking Scores on the Torrance Tests of Creative Thinking" *Creativity Research Journal*, 23:4, 285–295. Routledge, Taylor & Francis Group.
  https://www.tandfonline.com/doi/abs/10.1080/10400419.2011.627805, November 9, 2011.
  Downloaded by the College of William and Mary:
  https://kkim.wmwikis.net/file/view/Kim_2011_Creativity_crisis.pdf, April 3, 2012.

19

- Marie Winn. "Losing the Thread," "The Basic Building Blocks," "A Preference for Watching" *Plug-In Drug* (New York: Viking Penguin, 1977, Revised and Updated Edition, 2002) pp. 95–99.
- Herbert E. Krugman and Eugene L. Hartley. "Passive Learning From Television" *The Public Opinion Quarterly,* Vol. 34, No. 2. (Oxford University Press, 1970) pp. 184-190.
- Herbert E. Krugman. "Brain Wave Measures of Media Involvement" *How Advertising Works: The Role of Research.* (New York SAGE Publications, 1998) pp. 139–151.
- "Your brain waves change when you watch TV" *I Am Awake.* http://www.iamawake.co/your-brain-waves-change-when-you-watch-tv/, October 11, 2013.

20

- Kristina Birdsong. "This is Your Child's Brain on TV" *Scientific Learning.* http://www.scilearn.com/blog/how-television-impacts-learning, March 22, 2016.
  http://54.186.226.228/blog/your-childs-brain-tv, March 22, 2016.

21

- Rachael Pells. "Today's four-year-olds often 'not physically ready' for school, experts warn" *Independent.* https://www.independent.co.uk/news/education/education-news/school-age-four-year-old-children-not-physically-ready-experts-warn-a7220476.html, September 1, 2016.

22

- Aleeya Healey and Alan Mendelsohn. "Selecting Appropriate Toys for Young Children in the Digital Era" *American Academy of Pediatrics.* http://pediatrics.aappublications.org/content/early/2018/11/29/peds.2018-3348, December 2018.
  Print Source: PEDIATRICS Volume 143, number 1, January 2019 (Copyright © American Academy of Pediatrics 2019).

23

- Alexandra Thompson. “Surgery students are struggling to use their hands because they spent 'too much time watching TV' rather than sewing or playing an instrument as children” *The Daily Mail.* https://www.dailymail.co.uk/health/article-6332847/Surgery-students-struggle-use-hands-spent-time-watching-TV.html, October 30, 2018.

24

- Marie Winn. “A Commitment to Language” *Plug-In Drug.* (New York: Viking Penguin, 1977, Revised and Updated Edition, 2002) p. 76.
- Alice Park. “Baby Einsteins: Not So Smart After All” *Time.* http://content.time.com/time/health/article/0,8599,1650352,00.htm, August 06, 2007.
- Meta van den Heuvel, Julia Ma, Cornelia M Borkhoff, Christine Koroshegyi, David W H Dai, Patricia C Parkin, Jonathon L Maguire, Catherine S Birken. “Mobile Media Device Use is Associated with Expressive Language Delay in 18-Month-Old Children.” *Journal of developmental and behavioral pediatrics : JDBP vol. 40,2 (2019): 99-104. doi:10.1097/DBP.0000000000000630.* https://pubmed.ncbi.nlm.nih.gov/30753173/, Feb/Mar 2019.
- Sheri Madigan, PhD; Dillon Browne, PhD; Nicole Racine, PhD. “Association Between Screen Time and Children’s Performance on a Developmental Screening Test” *American Medical Association: JAMA Pediatr. 2019;173(3):244-250. doi:10.1001/jamapediatrics.2018.5056.* https://jamanetwork.com/journals/jamapediatrics/fullarticle/2722666, January 28, 2019.

25

- David L. Hill. “Why to Avoid TV Before Age 2” (early brain development). *American Academy of Pediatrics.* http://www.healthychildren.org/English/family-life/Media/Pages/Why-to-Avoid-TV-Before-Age-2.aspx, May 11, 2013.
  Original Source: Dad to Dad: Parenting Like a Pro (Copyright © American Academy of Pediatrics 2012).
- Alice Park. “Baby Einsteins: Not So Smart After All” *Time.* http://content.time.com/time/health/article/0,8599,1650352,00.html August 06, 2007.

- "Television vs. Reading" *Parent Soup®*, a Trademark of iVillage℠ Inc. Copyright 1996. http://webshare.northseattle.edu/fam180/topics/tv/tvvsread.htm, accessed May 29, 2014.
- Jim Trelease. *The Read-Aloud Handbook*, Penguin Books, 1995.
- Marie Winn. "Television and the Brain" and "Brain Changes" *Plug-In Drug.* (New York: Viking Penguin, 1977, Revised and Updated Edition, 2002) pp. 67–69.

26

- Carey Bryson. "Babies and TV: Is Screen Time Good for Your Little One?" *ThoughtCo.* https://www.thoughtco.com/should-babies-watch-tv-2107982, June 22, 2017.

27

- E. Tory Higgins. "Self-Discrepancy Theory" *Advances in Experimental Social Psychology*, Volume 22. Leonard Berkowitz, ed. (Cambridge, Massachusetts: Academic Press. March 28, 1989) p. 119.

28

- E. Tory Higgins. "Self-Discrepancy Theory" *Advances in Experimental Social Psychology*, Volume 22. Leonard Berkowitz, ed. (Cambridge, Massachusetts: Academic Press. March 28, 1989) p. 118.

29

- Nielsen. "Percentage of Americans who say they watch too much TV: 49 %" *BLS American Time Use Survey*, A.C. Nielsen Co. http://www.statisticbrain.com/television-watching-statistics/, Date Verified: 12.7.2013.
- Robert Kubey and Mihaly Csikszentmihalyi "Television Addiction Is No Mere Metaphor" *Scientific American*. http://www.academia.edu/5065840/Television_Addiction_is_no_mere_metaphor, January 23, 2016.

30

- E. Tory Higgins. "Self-Discrepancy Theory" *Advances in Experimental Social Psychology*, Volume 22. Leonard Berkowitz, ed. (Cambridge, Massachusetts: Academic Press. March 28, 1989) p. 118.

31

- Jeff Haden. *The Motivation Myth.* p. 110. New York: Penguin, 2018.

32

- Emma Henderson. "Watching lots of TV 'makes you stupid'" *The Independent.* http://www.independent.co.uk/news/science/watching-lots-of-tv-makes-you-stupid-says-american-universities-a6759026.html, December 3, 2015.
- Tina D. Hoang, MSPH; Jared Reis, PhD; Na Zhu, MD, MPH; David R. Jacobs Jr, PhD; Lenore J. Launer, PhD; Rachel A. Whitmer, PhD; Stephen Sidney, MD; Kristine Yaffe, MD. "Effect of Early Adult Patterns of Physical Activity and Television Viewing on Midlife Cognitive Function" *JAMA Psychiatry* (The Journal of the American Medical Association). http://archpsyc.jamanetwork.com/article.aspx?articleid=2471270, January 2016, Vol 73, No. 1.

33

- Douglas Fields. "Watching TV Alters Children's Brain Structure and Lowers IQ". http://rdouglasfields.com/2015/05/watching-tv-alters-childrens-brain-structure-and-lowers-iq/, May 4, 2015.

34

- Azmaira H. Maker. "Screen Time: The Impact on Kids and Parenting: New research explains the significant negative effects of excessive screen time." *Psychology Today.* https://www.psychologytoday.com/us/blog/helping-kids-cope/201808/screen-time-the-impact-kids-and-parenting, August 19, 2018.

35

- Victoria L. Dunckley, MD. Video Talk (*YouTube*). https://youtu.be/SWQuNnCdN5Y, November 8, 2018.
- Victoria L. Dunckley, MD. "Gray Matters: Too Much Screen Time Damages the Brain" *Psychology Today.* https://www.psychologytoday.com/us/blog/mental-wealth/201402/gray-matters-too-much-screen-time-damages-the-brain, February 27, 2014.

36

- Sandee LaMotte. "MRIs show screen time linked to lower brain development in preschoolers" *CNN*. https://www.cnn.com/2019/11/04/health/screen-time-lower-brain-development-preschoolers-wellness/index.html, November 4, 2019.
- John S. Hutton, MS, MD; Jonathan Dudley, PhD; Tzipi Horowitz-Kraus, PhD. "Associations Between Screen-Based Media Use and Brain White Matter Integrity in Preschool-Aged Children" *American Medical Association: JAMA Pediatr. 2020;174(1):e193869. doi:10.1001/jamapediatrics.2019.3869*. https://jamanetwork.com/journals/jamapediatrics/fullarticle/2754101, November 4, 2019.

37

- Maryanne Wolf. "Skim reading is the new normal. The effect on society is profound" *The Guardian*. https://www.theguardian.com/commentisfree/2018/aug/25/skim-reading-new-normal-maryanne-wolf, August 25, 2018.

38

- Maryanne Wolf. "Skim reading is the new normal. The effect on society is profound" *The Guardian*. https://www.theguardian.com/commentisfree/2018/aug/25/skim-reading-new-normal-maryanne-wolf, August 25, 2018.

39

- Alice Cuddy. "The IQ of Europeans is dropping due to technology, say researchers" *Euronews*. https://www.euronews.com/2017/12/29/the-iq-of-europeans-is-dropping-due-to-technology-say-researchers, December 29, 2017.

40

- James Flynn, Michael Shayer. "IQ decline and Piaget: Does the rot start at the top?" *Intelligence,* Volume 66, Pages 112-121. Elsevier B.V. https://doi.org/10.1016/j.intell.2017.11.010, December 8, 2017.

41

- Justin Kruger and David Dunning. "Unskilled and Unaware of It: How Difficulties in Recognizing One's Own Incompetence Lead to Inflated Self-Assessments" *Journal of Personality and Social Psychology.* 1999, Vol. 77, No. 6, 1121-1134. American Psychological Association. https://psycnet.apa.org/doiLanding?doi=10.1037%2F0022-3514.77.6.1121

42

- David Dunning. "We Are All Confident Idiots" *Pacific Standard.* https://www3.nd.edu/~ghaeffel/ConfidentIdiots.pdf, October 27, 2014.

43

- Tom Nichols. *The Death of Expertise.* p. 132. New York: Oxford University Press, 2017.

44

- Thomas Friedman. *Thank You for Being Late.* p. 378. New York: Farrar, Straus and Giroux (Picador), 2017.

45

- Shahram Heshmat. "What Is Confirmation Bias?" *Psychology Today.* https://www.psychologytoday.com/blog/science-choice/201504/what-is-confirmation-bias, April 23, 2015.

46

- Tom Nichols. *The Death of Expertise.* pp. 140, 155. New York: Oxford University Press, 2017.

47

- Jeff Grabmeier. "Both liberals, conservatives can have science bias: Study finds different topics bedevil the left and right" *The Ohio State University.* https://news.osu.edu/news/2015/02/09/both-liberals-conservatives-can-have-science-bias/, February 09, 2015.

- Chris Mooney. “Liberals deny science, too” *The Washington Post.* https://www.washingtonpost.com/news/wonk/wp/2014/10/28/liberals-deny-science-too/, October 28, 2014.
- Tom Nichols. *The Death of Expertise.* p. 69. New York: Oxford University Press, 2017.

48

- Tom Nichols. *The Death of Expertise.* pp. 119–122, 138. New York: Oxford University Press, 2017.
- Matthew Fisher et al. “Searching for Explanations: How the Internet Inflates Estimates of Internal Knowledge” *Journal of Experimental Psychology* 144(3). pp. 674–687, June 2015.
- Tom Jacobs. “Searching the Internet Creates an Illusion of Knowledge” *Pacific Standard online.* https://psmag.com/environment/searching-internet-creates-the-illusion-of-knowledge-, April 1, 2015.
- Richard Arum. “College Graduates: Satisfied, but Adrift” in *The State of the American Mind.* p. 73, Mark Bauerlein and Adam Bellow, eds. West Conshohocken, PA: Templeton, 2015.

49

- Tom Nichols. *The Death of Expertise.* pp. 119–122, 138. New York: Oxford University Press, 2017.
- Matthew Fisher et al. “Searching for Explanations: How the Internet Inflates Estimates of Internal Knowledge” *Journal of Experimental Psychology* 144(3). pp. 674–687, June 2015.
- Tom Jacobs. “Searching the Internet Creates an Illusion of Knowledge” *Pacific Standard online.* https://psmag.com/environment/searching-internet-creates-the-illusion-of-knowledge-, April 1, 2015.
- Richard Arum. “College Graduates: Satisfied, but Adrift” in *The State of the American Mind.* p. 73, Mark Bauerlein and Adam Bellow, eds. West Conshohocken, PA: Templeton, 2015.

50

- Joe Keohane. "How Facts Backfire: Researchers Discover a Surprising Threat to Democracy: Our Brains" Boston Globe online. http://archive.boston.com/news/science/articles/2010/07/11/how_facts_backfire/, July 11, 2010.
- Tom Nichols. *The Death of Expertise.* p. 131. New York: Oxford University Press, 2017.

51

- Tom Nichols. *The Death of Expertise.* p. 69. New York: Oxford University Press, 2017.

52

- Kyung Hee Kim. "The Creativity Crisis: The Decrease in Creative Thinking Scores on the Torrance Tests of Creative Thinking" *Creativity Research Journal*, 23:4, 285–295. Routledge, Taylor & Francis Group. https://www.tandfonline.com/doi/abs/10.1080/10400419.2011.627805, November 9, 2011.
  Downloaded by the College of William and Mary: https://kkim.wmwikis.net/file/view/Kim_2011_Creativity_crisis.pdf, April 3, 2012.
  TTCT-Figural Information:
  - "The *Torrance Tests of Creative Thinking* (TTCT) was developed in 1966 and renormed five times: in 1974, 1984, 1990, 1998, and 2008. The total sample for all six normative samples included 272,599 kindergarten through 12th grade students and adults."
  - "The TTCT has been translated into over 35 languages and it is utilized worldwide."
  - "The TTCT-Figural has been found to be fair in terms of gender, race, community status, language background, socioeconomic status, and culture."

53

- Kyung Hee Kim. "The Creativity Crisis: The Decrease in Creative Thinking Scores on the Torrance Tests of Creative Thinking" *Creativity Research Journal*, 23:4, 285–295. Routledge, Taylor & Francis Group. https://www.tandfonline.com/doi/abs/10.1080/10400419.2011.627805, November 9, 2011.
  Downloaded by the College of William and Mary: https://kkim.wmwikis.net/file/view/Kim_2011_Creativity_crisis.pdf, April 3, 2012.

54

- Jean M. Twenge, W. Keith Campbell. "Associations between screen time and lower psychological well-being among children and adolescents: Evidence from a population-based study" *Preventive Medicine Reports,* Volume 12, Pages 271-283. Elsevier B.V. https://doi.org/10.1016/j.pmedr.2018.10.003, October 18, 2018.

55

- Nellie Bowles. "Human Contact Is Now a Luxury Good: Screens used to be for the elite. Now avoiding them is a status symbol." *The New York Times.* https://www.nytimes.com/2019/03/23/sunday-review/human-contact-luxury-screens.html, March 23, 2019.
- Anderson Cooper. "Groundbreaking study examines effects of screen time on kids" *CBSNews.* https://www.cbsnews.com/news/groundbreaking-study-examines-effects-of-screen-time-on-kids-60-minutes/, December 9, 2018.

56

- Jessica Stillman. "Which Country Has the Most Productive Workers?" *Inc.com.* http://www.inc.com/jessica-stillman/which-country-has-the-most-productive-workers.html, July 25, 2014.

57

- Jessica Stillman. "The Cold, Hard Truth: You're Overwhelmed Because You Want to Be" *Inc.com.* http://www.inc.com/jessica-stillman/cure-for-feeling-overwhelmed-is-to-stop-talking-about-it.html, March 28, 2014.

- Brigid Schulte. *Overwhelmed: Work, Love, and Play When No One Has the Time.* New York: Simon & Schuster, Inc./Sarah Crichton Books, 2014.

58

- Stephen Johnson. "China's schoolkids beat American students in all academic categories" *bigthink.com*. https://bigthink.com/the-present/pisa-test-china/#rebelltitem4, December 5, 2019.
- Tatumn Walter. "A tenth of each day spent watching television" *USC US-China Institute*. https://china.usc.edu/tenth-each-day-spent-watching-television, accessed May 29, 2014.
  This study shows Americans watch 5 hours and 17 minutes a day. Call it 5 hours and 15 minutes (or 5.25 hours / day), and your first 22 years lost 42,157 hours of development time. This study also notes American TV-watching is more than double that of the Chinese.
- Arjun Kharpal, Lauren Feiner. "China to ban kids from playing online games for more than three hours per week" *CNBC*. https://cnb.cx/2V4tkMw
  (also https://www.cnbc.com/2021/08/30/china-to-ban-kids-from-playing-online-games-for-more-than-three-hours-per-week.html), August 30 2021.

59

- Travis Bradberry and Jean Greaves. *Emotional Intelligence 2.0.* p. 242. San Diego, California: TalentSmart, 2009.

60

- Jamil Zaki. "What, Me Care? Young Are Less Empathetic" *Scientific American*. A recent study finds a decline in empathy among young people in the U.S. http://www.scientificamerican.com/article/what-me-care/, December 23, 2010.

- "Children who watch 'excessive' amounts of TV are more likely to have criminal convictions, exhibit aggression and experience negative emotions: study." *New York Daily News.* http://www.nydailynews.com/life-style/health/kids-watch-excessive-tv-criminal-convictions-young-adulthood-study-article-1.1267868, February 19, 2013.
  "With every hour in front of the television, kids were more likely to show aggressive behavior or receive a criminal conviction by early adulthood, according to a study published in 'Pediatrics.' The issue isn't necessarily the content of the programming, but the social isolation that comes from so many hours in front of the tube."
- "TV retards your child's development" *Consumers Association of Penang.* http://www.consumer.org.my/index.php/development/education/347-tv-retards-your-childs-development, accessed July 17, 2016.
  Also See:
  "More arguments against TV" Consumers Association of Penang. https://consumer.org.my/more-arguments-against-tv/, March 18, 2022.

61

- Marie Winn. "Sesame Street Revisited" *Plug-In Drug.* pp. 60, 61. Viking Penguin, 1977, Revised and Updated Edition, 2002.
- Alice Park. "Baby Einsteins: Not So Smart After All" *Time.* http://content.time.com/time/health/article/0,8599,1650352,00.html August 06, 2007.
- *New York Daily News.* Link: http://www.nydailynews.com/life-style/health/kids-watch-excessive-tv-criminal-convictions-young-adulthood-study-article-1.1267868, February 19, 2013.
  "Children who watch 'excessive' amounts of TV are more likely to have criminal convictions, exhibit aggression and experience negative emotions: study. With every hour in front of the television, kids were more likely to show aggressive behavior or receive a criminal conviction by early adulthood, according to a study published in 'Pediatrics.' The issue isn't necessarily the content of the programming, but the social isolation that comes from so many hours in front of the tube."

62

- "TV retards your child's development" *Consumers Association of Penang.* http://www.consumer.org.my/index.php/development/education/347-tv-retards-your-childs-development, accessed July 17, 2016.
  Also See:
  "More arguments against TV" Consumers Association of Penang. https://consumer.org.my/more-arguments-against-tv/, March 18, 2022.

63

- Marie Winn. "Sesame Street Revisited" *Plug-In Drug.* pp. 60, 61. Viking Penguin, 1977, Revised and Updated Edition, 2002.
- Alice Park. "Baby Einsteins: Not So Smart After All" *Time*. http://content.time.com/time/health/article/0,8599,1650352,00.html August 06, 2007.
- *New York Daily News*. Link: http://www.nydailynews.com/life-style/health/kids-watch-excessive-tv-criminal-convictions-young-adulthood-study-article-1.1267868, February 19, 2013.
  "Children who watch 'excessive' amounts of TV are more likely to have criminal convictions, exhibit aggression and experience negative emotions: study. With every hour in front of the television, kids were more likely to show aggressive behavior or receive a criminal conviction by early adulthood, according to a study published in 'Pediatrics.' The issue isn't necessarily the content of the programming, but the social isolation that comes from so many hours in front of the tube."

64

- Herbert E. Krugman and Eugene L. Hartley. "Passive Learning From Television" *The Public Opinion Quarterly,* Vol. 34, No. 2. (Oxford University Press, 1970) pp. 184-190.
- Herbert E. Krugman. "Brain Wave Measures of Media Involvement" *How Advertising Works: The Role of Research.* (New York SAGE Publications, 1998) pp. 139–151.
- "Your brain waves change when you watch TV" *I Am Awake.* http://www.iamawake.co/your-brain-waves-change-when-you-watch-tv/, October 11, 2013.

65

- Tom Nichols. *The Death of Expertise.* p. 69. New York: Oxford University Press, 2017.

66

- Justin Kruger and David Dunning. "Unskilled and Unaware of It: How Difficulties in Recognizing One's Own Incompetence Lead to Inflated Self-Assessments" *Journal of Personality and Social Psychology.* 1999, Vol. 77, No. 6, 1121-1134. American Psychological Association. https://psycnet.apa.org/doiLanding?doi=10.1037%2F0022-3514.77.6.1121

67

- Allan Stromfeldt Christensen. "Lemminged: to be herded off the peak oil cliff by filmmakers" *TransitionVoice.com.* http://transitionvoice.com/2014/11/lemminged-to-be-herded-off-the-peak-oil-cliff-by-filmmakers/, November 11, 2014.

68

- Robert Kubey and Mihaly Csikszentmihalyi. "Television Addiction is no mere metaphor" *Scientific American.* http://www.academia.edu/5065840/Television_Addiction_is_no_mere_metaphor, accessed January 23, 2016.

69

- David Hinckley. "Average American watches 5 hours of TV per day, report shows" *New York Daily News.* https://www.nydailynews.com/life-style/average-american-watches-5-hours-tv-day-article-1.1711954, March 5, 2014.
  Per Nielsen, here's the average weekly usage for ascending age groups. Ages:
  - 2–11: 24 hours, 16 minutes. (round down to 24 hours, 15 minutes, or 24.25 x 10 years counting ages 2 and 11 = 12,610)
  - 12–17: 20 hours, 41 minutes. (round down to 20 hours, 30 minutes, or 20.5 hours x 6 years counting ages 12 and 17 = 6,396)
  - 18–24: 22 hours, 27 minutes. (round down to 22 hours, 20 minutes, or 20.33 hours x 6 years counting age 18 to end of age 23 = 6,343)

- By age 24, the average American has watched 25,349 hours of TV (does not include Time on other devices). Here is the breakdown continuing older age groups, showing people watch more and more as they get older. Ages:
  - 25–34: 27 hours, 36 minutes.
  - 35–49: 33 hours, 40 minutes.
  - 50–64: 43 hours, 56 minutes.
  - 65–plus: 50 hours, 34 minutes.

70

- Marie Winn. "Losing the Thread," "The Basic Building Blocks," "A Preference for Watching" *Plug-In Drug.* (New York: Viking Penguin, 1977, Revised and Updated Edition, 2002) pp. 95–99.

71

- Kristina Birdsong. "This is Your Child's Brain on TV" *Scientific Learning.* http://www.scilearn.com/blog/how-television-impacts-learning, March 22, 2016.
  http://54.186.226.228/blog/your-childs-brain-tv, March 22, 2016.

72

- Josh Constine. "Jeff Bezos' guide to life" *TechCrunch.* https://techcrunch.com/2017/11/05/jeff-bezos-guide-to-life/, November 5, 2017.

73

- Jon Hamilton. "How Play Wires Kids' Brains For Social and Academic Success" *KQED News.* KQED.org, National Public Radio (NPR), Copyright 2014.
  http://ww2.kqed.org/mindshift/2014/08/07/how-play-wires-kids-brains-for-social-and-academic-success/, August 7, 2014.

74

- Lost development time could be more than 40,000 hours lost in 22 years, counting passive video time on all devices. The number from earlier estimates cited in this book is a conservative 25,349 by age 24. Using other statistics:
  - Nielsen's average hours breakdown for American youth, Americans watch 26,400 hours in their first 22 years.

Nielsen. BLS American Time Use Survey, A.C. Nielsen Co. http://www.statisticbrain.com/television-watching-statistics/, Date Verified: 12.7.2013 (July 12, 2013).

- Counting 3 devices—TV video, PC-Internet video, mobile-phone video—Average American viewing time is 151 hours per month = 39,864 in 22 years.
  Nielsen. "Television, Internet and Mobile Usage in the U.S. — A2/M2 Three Screen Report 4th Quarter 2008". Copyright © 2009 The Nielsen Company.
  http://i.cdn.turner.com/cnn/2009/images/02/24/screen.press.b.pdf, accessed May 29, 2014.
- Another study shows Americans watch 5 hours and 17 minutes a day. Call it 5 hours and 15 minutes (or 5.25 hours / day), then your first 22 years lost 42,157 hours of development time.
  Tatumn Walter. "A tenth of each day spent watching television" *USC US-China Institute*.
  https://china.usc.edu/tenth-each-day-spent-watching-television, accessed May 29, 2014.
- Still another summary of TV studies says "When other screen-based viewing, such as computer games, is included, the figure is far higher. Children aged 11 to 15 now spend 53 hours a week watching TV and computers"
  Aric Sigman. "How TV Is (Quite Literally) Killing Us" *Daily Mail*.
  http://www.whale.to/b/sigman.html, Oct 1, 2005.

- This amount of lost development is leading to more and more video brain-dependency passed along to new generations so that higher brain function will disappear completely.

75

- Victor C. Strasburger, MD, FAAP, and Marjorie J. Hogan, MD, FAAP. "Children, Adolescents, and the Media" *American Academy of Pediatrics*. http://pediatrics.aappublications.org/content/132/5/958.full, May 11, 2013.

76

- "TV retards your child's development" *Consumers Association of Penang.* http://www.consumer.org.my/index.php/development/education/347-tv-retards-your-childs-development, accessed July 17, 2016.
  Also See:

"More arguments against TV" Consumers Association of Penang. https://consumer.org.my/more-arguments-against-tv/, March 18, 2022.

- Emma Henderson. "Watching lots of TV 'makes you stupid'" *The Independent.* http://www.independent.co.uk/news/science/watching-lots-of-tv-makes-you-stupid-says-american-universities-a6759026.html, December 3, 2015.
- Herbert E. Krugman and Eugene L. Hartley. "Passive Learning From Television" *The Public Opinion Quarterly,* Vol. 34, No. 2. (Oxford University Press, 1970) pp. 184-190.
- Herbert E. Krugman. "Brain Wave Measures of Media Involvement" *How Advertising Works: The Role of Research.* (New York SAGE Publications, 1998) pp. 139–151.
- "Your brain waves change when you watch TV" *I Am Awake.* http://www.iamawake.co/your-brain-waves-change-when-you-watch-tv/, October 11, 2013.
- Kristina Birdsong. "This is Your Child's Brain on TV" *Scientific Learning.* http://www.scilearn.com/blog/how-television-impacts-learning, March 22, 2016.
  http://54.186.226.228/blog/your-childs-brain-tv, March 22, 2016.
- Douglas Fields. "Watching TV Alters Children's Brain Structure and Lowers IQ". http://rdouglasfields.com/2015/05/watching-tv-alters-childrens-brain-structure-and-lowers-iq/, May 4, 2015.
- Jamil Zaki. "What, Me Care? Young Are Less Empathetic" *Scientific American*. A recent study finds a decline in empathy among young people in the U.S. http://www.scientificamerican.com/article/what-me-care/, December 23, 2010.
- "Children who watch 'excessive' amounts of TV are more likely to have criminal convictions, exhibit aggression and experience negative emotions: study." *New York Daily News*. http://www.nydailynews.com/life-style/health/kids-watch-excessive-tv-criminal-convictions-young-adulthood-study-article-1.1267868, February 19, 2013.
  "With every hour in front of the television, kids were more likely to show aggressive behavior or receive a criminal conviction by early adulthood, according to a study published in 'Pediatrics.' The issue isn't necessarily the content of the programming, but the social isolation that comes from so many hours in front of the tube."

- David L. Hill. "Why to Avoid TV Before Age 2" (early brain development). *American Academy of Pediatrics*.
  http://www.healthychildren.org/English/family-life/Media/Pages/Why-to-Avoid-TV-Before-Age-2.aspx, May 11, 2013.
  Original Source: Dad to Dad: Parenting Like a Pro (Copyright © American Academy of Pediatrics 2012).
- Alice Park. "Baby Einsteins: Not So Smart After All" *Time*.
  http://content.time.com/time/health/article/0,8599,1650352,00.html August 06, 2007.
- "Television vs. Reading" *Parent Soup*®, a Trademark of iVillage℠ Inc. Copyright 1996.
  http://webshare.northseattle.edu/fam180/topics/tv/tvvsread.htm, accessed May 29, 2014.
- Jim Trelease. *The Read-Aloud Handbook*, Penguin Books, 1995.
- Edward L. Schor. *Caring for your school-age child : ages 5 to 12.* American Academy of Pediatrics. New York : Bantam Books, 2004.
  "What Children are NOT Doing When Watching TV" American Academy of Pediatrics.
  http://www.healthychildren.org/English/family-life/Media/Pages/What-Children-are-NOT-Doing-When-Watching-TV.aspx, May 11, 2013.
  Also See: *Caring for Our Children*. A Joint Collaborative Project of the American Academy of Pediatrics and the American Public Health Association.
  https://nrckids.org/files/CFOC4%20pdf-%20FINAL.pdf, 2019.
- Tina D. Hoang, MSPH; Jared Reis, PhD; Na Zhu, MD, MPH; David R. Jacobs Jr, PhD; Lenore J. Launer, PhD; Rachel A. Whitmer, PhD; Stephen Sidney, MD; Kristine Yaffe, MD.
  "Effect of Early Adult Patterns of Physical Activity and Television Viewing on Midlife Cognitive Function" *JAMA Psychiatry* (The Journal of the American Medical Association).
  http://archpsyc.jamanetwork.com/article.aspx?articleid=2471270, January 2016, Vol 73, No. 1.
- Marie Winn. "Television and the Brain," "Brain Changes," "Losing the Thread," "The Basic Building Blocks," "A Preference for Watching," "Free Time and Resourcefulness" *Plug-In Drug.* (New York: Viking Penguin, 1977, Revised and Updated Edition, 2002) pp. 67–69, 95–99, 131.

77

- Victoria L. Dunckley, MD. "Dumb & Dumber: Interactive Screentime is Worse than TV" *Psychology Today.* https://www.psychologytoday.com/us/blog/mental-wealth/201408/dumb-dumber-interactive-screentime-is-worse-tv, August 29, 2014.

78

- Robert Kubey and Mihaly Csikszentmihalyi. "Television Addiction is no mere metaphor" *Scientific American.* http://www.academia.edu/5065840/Television_Addiction_is_no_mere_metaphor, accessed January 23, 2016.

79

- Robert Kubey and Mihaly Csikszentmihalyi. "Television Addiction is no mere metaphor" *Scientific American.* http://www.academia.edu/5065840/Television_Addiction_is_no_mere_metaphor, accessed January 23, 2016.

80

- Bruno S. Frey, Christine Benesch, Alois Stutzer. "Does watching TV make us happy?" *Journal of Economic Psychology*, Volume 28, Issue 3, June 2007. Elsevier B.V. (ScienceDirect.com). https://www.bsfrey.ch/wp-content/uploads/2021/05/does-watching-tv-make-us-happy.pdf, February 14, 2007.

81

- Guangheng Dong, Yanbo Hu, Xiao Lin. "Reward/punishment sensitivities among internet addicts: Implications for their addictive behaviors" Progress in Neuro-Psychopharmacology and Biological Psychiatry, Volume 46, October 1, 2013, Pages 139-145. https://www.sciencedirect.com/science/article/pii/S0278584613001486, July 19, 2013.
- Kai Yuan, Ping Cheng, Tao Dong, et al. "Cortical Thickness Abnormalities in Late Adolescence with Online Gaming Addiction" PLOS ONE. https://doi.org/10.1371/journal.pone.0053055, January 9, 2013.
- Kai Yuan, Wei Qin, Guihong Wang, et al. "Microstructure Abnormalities in Adolescents with Internet Addiction Disorder" PLOS ONE. https://doi.org/10.1371/journal.pone.0020708, June 3, 2011.

82

- "Most teenagers know smartphones compromise their lives, wish to curb use" *The Economic Times.* https://economictimes.indiatimes.com/magazines/panache/most-teenagers-know-smartphones-compromise-their-lives-wish-to-curb-use/articleshow/64707105.cms, June 23, 2018.
  Also see:
  Annie Sneed. "Are you addicted to your phone? Here is how to cut back" The Economic Times.
  https://economictimes.indiatimes.com/news/how-to/are-you-addicted-to-your-phone-here-is-how-you-can-cut-back/articleshow/89607023.cms, February 16, 2022.

83

- Richard Freed. "The Tech Industry's War on Kids: How psychology is being used as a weapon against children" *Blog, Child and adolescent psychologist* Richard Freed.
  https://medium.com/@richardnfreed/the-tech-industrys-psychological-war-on-kids-c452870464ce, March 12, 2018.

84

- Hilary Andersson. "Social media apps are 'deliberately' addictive to users" *BBC.*
  https://www.bbc.com/news/technology-44640959, July 4, 2018.

85

- Hilary Andersson. "Social media apps are 'deliberately' addictive to users" *BBC.*
  https://www.bbc.com/news/technology-44640959, July 4, 2018.

86

- Georgia Wells, Jeff Horwitz, and Deepa Seetharaman. "Facebook Knows Instagram Is Toxic for Teen Girls, Company Documents Show" *The Wall Street Journal.*
  https://www.wsj.com/articles/facebook-knows-instagram-is-toxic-for-teen-girls-company-documents-show-11631620739, September 14, 2021.

87

- Georgia Wells, Jeff Horwitz, and Deepa Seetharaman. “Facebook Knows Instagram Is Toxic for Teen Girls, Company Documents Show” *The Wall Street Journal.* https://www.wsj.com/articles/facebook-knows-instagram-is-toxic-for-teen-girls-company-documents-show-11631620739, September 14, 2021.

88

- Adam Alter. “Tech Bigwigs Know How Addictive Their Products Are. Why Don’t the Rest of Us?” *Wired.com.* https://www.wired.com/2017/03/irresistible-the-rise-of-addictive-technology-and-the-business-of-keeping-us-hooked/, March 24, 2017.

89

- Mansoor Iqbal. “TikTok Revenue and Usage Statistics (2022)” *Business of Apps*. https://www.businessofapps.com/data/tik-tok-statistics/, November 11, 2022.
- David Curry. “BeReal Revenue and Usage Statistics (2022)” *Business of Apps*. https://www.businessofapps.com/data/bereal-statistics/, November 10, 2022.
- Mansoor Iqbal. “Snapchat Revenue and Usage Statistics (2022)” *Business of Apps*. https://www.businessofapps.com/data/snapchat-statistics/, August 31, 2022.
- Mansoor Iqbal. “Facebook Revenue and Usage Statistics (2022)” *Business of Apps*. https://www.businessofapps.com/data/facebook-statistics/, November 24, 2022.
- Mansoor Iqbal. “Instagram Revenue and Usage Statistics (2022)” *Business of Apps*. https://www.businessofapps.com/data/instagram-statistics/, September 6, 2022.
- Mansoor Iqbal. “Twitter Revenue and Usage Statistics (2022)” *Business of Apps*. https://www.businessofapps.com/data/twitter-statistics/, November 4, 2022.

90

- John Koetsier. “Digital Crack Cocaine: The Science Behind TikTok’s Success” *Forbes*. https://www.forbes.com/sites/johnkoetsier/2020/01/18/digital-crack-cocaine-the-science-behind-tiktoks-success/, January 18, 2020.

91

- Rebecca R. Ruiz. “Betting Apps Can Make Anyone a Sports Fan. Even Me” *The New York Times*. https://www.nytimes.com/2022/12/24/business/sports-betting-apps.html, December 24, 2022.

92

- Rebecca R. Ruiz. “Betting Apps Can Make Anyone a Sports Fan. Even Me” *The New York Times*. https://www.nytimes.com/2022/12/24/business/sports-betting-apps.html, December 24, 2022.

93

- John Koetsier. “Digital Crack Cocaine: The Science Behind TikTok’s Success” *Forbes*. https://www.forbes.com/sites/johnkoetsier/2020/01/18/digital-crack-cocaine-the-science-behind-tiktoks-success/, January 18, 2020.
- Sophia Petrillo. “What Makes TikTok so Addictive?: An Analysis of the Mechanisms Underlying the World’s Latest Social Media Craze” *Brown Undergraduate Journal of Public Health*. https://sites.brown.edu/publichealthjournal/2021/12/13/tiktok/, December 13, 2021.
- R. Burhan, J. Moradzadeh. “Neurotransmitter Dopamine (DA) and its Role in the Development of Social Media Addiction” *Journal of Neurology*, *11*(7), 507. https://www.iomcworld.org/open-access/neurotransmitter-dopamine-da-and-its-role-in-the-development-of-social-media-addiction-59222.html, 2020.
- R. J. Lee-Won, L. Herzog, S. G. Park. “Hooked on Facebook: The role of social anxiety and need for social assurance in problematic use of Facebook” *Cyberpsychology, Behavior, and Social Networking*, *18*(10), 567-574. https://pubmed.ncbi.nlm.nih.gov/26383178/, Sep 18, 2015.

- L. Y. Lin, J. E. Sidani, A. Shensa, A. Radovic, E. Miller, J. B. Colditz, B. A. Primack. "Association between social media use and depression among U.S. young adults" *Depression and anxiety, 33*(4), 323-331. https://doi.org/10.1002/da.22466, January 19, 2016.
- Richard Fang. The psychology of why social media is so addictive. *UX Collective*. https://uxdesign.cc/the-psychology-of-why-social-media-is-so-addictive-67830266657d, September 21, 2020.
- K. Sriwilai, P. Charoensukmongkol. "Face it, don't Facebook it: impacts of social media addiction on mindfulness, coping strategies and the consequence on emotional exhaustion" *Stress and Health, 32*(4), 427-434. https://doi.org/10.1002/smi.2637, March 30, 2015.
- C. Montag, B. Lachmann, M. Herrlich, K. Zweig. "Addictive Features of Social Media/Messenger Platforms and Freemium Games against the Background of Psychological and Economic Theories" *International journal of environmental research and public health, 16*(14), 2612 (2019). https://doi.org/10.3390/ijerph16142612, July 23, 2019.
- L. E. Sherman, A. A. Payton, L. M. Hernandez, P. M. Greenfield, M. Dapretto, "The Power of the Like in Adolescence: Effects of Peer Influence on Neural and Behavioral Responses to Social Media" *Psychological science, 27*(7), 1027–1035. https://doi.org/10.1177/0956797616645673, May 31, 2016.
- H. C. Woods, H. Scott. "# Sleepyteens: Social media use in adolescence is associated with poor sleep quality, anxiety, depression and low self-esteem" *Journal of adolescence, 51*, 41-49. https://doi.org/10.1016/j.adolescence.2016.05.008, June 10, 2016.

94

- Marc Atherton. "Silicon Valley's Secret Sauce" *Swipe Left For Addiction*. https://swipeleftforaddiction.com/2018/05/30/silicon-valleys-secret-sauce/, accessed July 28, 2018.
  Updated source site:
  "The Attention Economy, Internet Addiction and Unintended Consequences" *Swipe Left For Addiction.* https://swipeleftforaddiction.wordpress.com/2018/05/11/the-attention-economy-internet-addiction-and-unintended-consequences/, May 11, 2018.

95

- Adam Alter. "Tech Bigwigs Know How Addictive Their Products Are. Why Don't the Rest of Us?" *Wired.com.* https://www.wired.com/2017/03/irresistible-the-rise-of-addictive-technology-and-the-business-of-keeping-us-hooked/, March 24, 2017.

96

- Adam Alter. "Tech Bigwigs Know How Addictive Their Products Are. Why Don't the Rest of Us?" *Wired.com.* https://www.wired.com/2017/03/irresistible-the-rise-of-addictive-technology-and-the-business-of-keeping-us-hooked/, March 24, 2017.

97

- Adam Alter. "Tech Bigwigs Know How Addictive Their Products Are. Why Don't the Rest of Us?" *Wired.com.* https://www.wired.com/2017/03/irresistible-the-rise-of-addictive-technology-and-the-business-of-keeping-us-hooked/, March 24, 2017.

98

- Joe Paul. "What is Nomophobia?" *TimeToLogOff.com*. https://www.itstimetologoff.com/2018/06/18/what-is-nomophobia/, June 2018.

99

- Joe Paul. "What is Nomophobia?" *TimeToLogOff.com*. https://www.itstimetologoff.com/2018/06/18/what-is-nomophobia/, June 2018.

100

- Joe Paul. "What is Nomophobia?" *TimeToLogOff.com*. https://www.itstimetologoff.com/2018/06/18/what-is-nomophobia/, June 2018.

101

- Mary Ann Liebert. "Understanding smartphone separation anxiety and what smartphones mean to people". *Phys.org.* https://phys.org/news/2017-08-smartphone-anxiety-smartphones-people.html, August 14, 2017.

- Seunghee Han; Jang Hyun Kim; Ki Joon Kim. “Understanding Nomophobia: Structural Equation Modeling and Semantic Network Analysis of Smartphone Separation Anxiety.” *US National Library of Medicine National Institutes of Health.* https://www.ncbi.nlm.nih.gov/pubmed/28650222, June 26, 2017.

102

- Nicholas Kardaras. “Kids turn violent as parents battle ‘digital heroin’ addiction” *The New York Post.* https://nypost.com/2016/08/27/its-digital-heroin-how-screens-turn-kids-into-psychotic-junkies/, December 17, 2016.

103

- Jim Taylor. “Pro Athletes are Hooked on Social Media Too”. https://www.drjimtaylor.com/4.0/pro-athletes-are-hooked-on-social-media-too/, March 26, 2018.
- Candace Buckner. “NBA players know they’re addicted to their phones. Good luck getting them to unplug” *The Washington Post.* https://www.washingtonpost.com/sports/nba-players-know-theyre-addicted-to-their-phones-good-luck-getting-them-to-unplug/2018/03/19/6165cb96-2563-11e8-b79d-f3d931db7f68_story.html, March 19, 2018.

104

- Richard Freed. “The Tech Industry’s War on Kids: How psychology is being used as a weapon against children” *Blog, Child and adolescent psychologist* Richard Freed. https://medium.com/@richardnfreed/the-tech-industrys-psychological-war-on-kids-c452870464ce, March 12, 2018.

105

- Richard Freed. “The Tech Industry’s War on Kids: How psychology is being used as a weapon against children” *Blog, Child and adolescent psychologist* Richard Freed. https://medium.com/@richardnfreed/the-tech-industrys-psychological-war-on-kids-c452870464ce, March 12, 2018.

106

- Marie Winn. “A Chilling Episode: The ‘Tired-Child Syndrome’” *Plug-In Drug.* (New York: Viking Penguin, 1977, Revised and Updated Edition, 2002) p.219, 220.

107

- Marie Winn. “A Chilling Episode: The ‘Tired-Child Syndrome’” *Plug-In Drug.* (New York: Viking Penguin, 1977, Revised and Updated Edition, 2002) p.219, 220.

108

- Adam Alter. “Tech Bigwigs Know How Addictive Their Products Are. Why Don’t the Rest of Us?” *Wired.com.* https://www.wired.com/2017/03/irresistible-the-rise-of-addictive-technology-and-the-business-of-keeping-us-hooked/, March 24, 2017.

109

- Chris Weller. “An MIT psychologist explains why so many tech moguls send their kids to anti-tech schools” *Business Insider.* https://www.businessinsider.com/sherry-turkle-why-tech-moguls-send-their-kids-to-anti-tech-schools-2017-11, November 7, 2017.

110

- Chris Weller. “An MIT psychologist explains why so many tech moguls send their kids to anti-tech schools” *Business Insider.* https://www.businessinsider.com/sherry-turkle-why-tech-moguls-send-their-kids-to-anti-tech-schools-2017-11, November 7, 2017.

111

- Adam Alter. “Why Our Screens Make Us Less Happy” *TED Conference.* https://www.ted.com/talks/adam_alter_why_our_screens_make_us_less_happy, April 2017.

112

- Matt Richtel. “A Silicon Valley School That Doesn’t Compute” *The New York Times.* https://www.nytimes.com/2011/10/23/technology/at-waldorf-school-in-silicon-valley-technology-can-wait.html, October 22, 2011.

113

- Adam Alter. "Tech Bigwigs Know How Addictive Their Products Are. Why Don't the Rest of Us?" *Wired.com.* https://www.wired.com/2017/03/irresistible-the-rise-of-addictive-technology-and-the-business-of-keeping-us-hooked/, March 24, 2017.
- Chris Weller. "An MIT psychologist explains why so many tech moguls send their kids to anti-tech schools" *Business Insider.* https://www.businessinsider.com/sherry-turkle-why-tech-moguls-send-their-kids-to-anti-tech-schools-2017-11, November 7, 2017.

114

- Adam Alter. "Why Our Screens Make Us Less Happy" *TED Conference.* https://www.ted.com/talks/adam_alter_why_our_screens_make_us_less_happy, April 2017.

115

- Matt Richtel. "A Silicon Valley School That Doesn't Compute" *The New York Times.* https://www.nytimes.com/2011/10/23/technology/at-waldorf-school-in-silicon-valley-technology-can-wait.html, October 22, 2011.

116

- Owenz, Meghan. "The Rich Get Smart, The Poor Get Technology: The New Digital Divide in School Choice"*Screenfreeparenting.com*. https://www.screenfreeparenting.com/rich-get-smart-poor-get-technology-new-digital-divide-school-choice/, November 21, 2017.

117

- Nellie Bowles. "The Digital Gap Between Rich and Poor Kids Is Not What We Expected" *The New York Times.* https://www.nytimes.com/2018/10/26/style/digital-divide-screens-schools.html, October 26, 2018.

118

- Nellie Bowles. "Human Contact Is Now a Luxury Good: Screens used to be for the elite. Now avoiding them is a status symbol." *The New York Times.* https://www.nytimes.com/2019/03/23/sunday-review/human-contact-luxury-screens.html, March 23, 2019.

- Anderson Cooper. "Groundbreaking study examines effects of screen time on kids" *CBSNews.* https://www.cbsnews.com/news/groundbreaking-study-examines-effects-of-screen-time-on-kids-60-minutes/, December 9, 2018.

119

- Nellie Bowles. "The Digital Gap Between Rich and Poor Kids Is Not What We Expected" *The New York Times.* https://www.nytimes.com/2018/10/26/style/digital-divide-screens-schools.html, October 26, 2018.

120

- Cory Turner. "The Surgeon Who Became An Activist For Baby Talk" *NPR Ed.* https://www.npr.org/sections/ed/2015/09/14/437515492/the-surgeon-who-became-an-activist-for-baby-talk, September 14, 2015.

121

- Kristina Birdsong. "This is Your Child's Brain on TV" *Scientific Learning.* http://www.scilearn.com/blog/how-television-impacts-learning, March 22, 2016.
  http://54.186.226.228/blog/your-childs-brain-tv, March 22, 2016.

122

- Louis-Philippe Beland and Richard Murphy. "Ill Communication: Technology, distraction & student performance" *Labour Economics.* Elsevier B.V. https://www.sciencedirect.com/science/article/abs/pii/S0927537116300136, April 16, 2016.

123

- Nicholas Kardaras. "The case against screens in schools" *New York Post.* https://nypost.com/2016/12/17/the-case-against-screens-in-schools/, December 17, 2016.

124

- Owenz, Meghan. "The Rich Get Smart, The Poor Get Technology: The New Digital Divide in School Choice"*Screenfreeparenting.com*. https://www.screenfreeparenting.com/rich-get-smart-poor-get-technology-new-digital-divide-school-choice/, November 21, 2017.

125

- Georgia Wells, Jeff Horwitz, and Deepa Seetharaman. "Facebook Knows Instagram Is Toxic for Teen Girls, Company Documents Show" *The Wall Street Journal*. https://www.wsj.com/articles/facebook-knows-instagram-is-toxic-for-teen-girls-company-documents-show-11631620739, September 14, 2021.

126

- Jean M. Twenge. "Have Smartphones Destroyed a Generation?" *The Atlantic*. https://www.theatlantic.com/magazine/archive/2017/09/has-the-smartphone-destroyed-a-generation/534198/, September 2017.
- Jean M. Twenge. "This Is What Happy Teens Do. Hint: It doesn't involve their phones." *Psychology Today*. https://www.psychologytoday.com/us/blog/our-changing-culture/201808/is-what-happy-teens-do, August 31, 2018.

127

- Jill Anderson. "Smartphones, teens, and unhappiness" *The Harvard Gazette, Harvard University*. https://news.harvard.edu/gazette/story/2018/06/gse-phones-study/, June 20, 2018.
- Jean Twenge. "Changes in how we're spending our free time might explain the unhappiness epidemic" *PsyPost.org*. https://www.psypost.org/2018/08/changes-in-how-were-spending-our-free-time-might-explain-the-unhappiness-epidemic-51893, August 5, 2018.
- Jean M. Twenge, G. N. Martin, W. K. Campbell. "Decreases in psychological well-being among American adolescents after 2012 and links to screen time during the rise of smartphone technology." American Psychological Association Journal *Emotion*, 18(6), 765–780. https://doi.org/10.1037/emo0000403, 2018.

128

- Jean M. Twenge, G. N. Martin, W. K. Campbell. "Decreases in psychological well-being among American adolescents after 2012 and links to screen time during the rise of smartphone technology." American Psychological Association Journal *Emotion*, 18(6), 765–780. https://doi.org/10.1037/emo0000403, 2018.

129

- Jean M. Twenge. "Have Smartphones Destroyed a Generation?" *The Atlantic*. https://www.theatlantic.com/magazine/archive/2017/09/has-the-smartphone-destroyed-a-generation/534198/, September 2017.

130

- Jean M. Twenge, Thomas E. Joiner, Megan L. Rogers, Gabrielle N. Martin. "Increases in Depressive Symptoms, Suicide-Related Outcomes, and Suicide Rates Among U.S. Adolescents After 2010 and Links to Increased New Media Screen Time" *Clinical Psychological Science*. http://journals.sagepub.com/doi/full/10.1177/2167702617723376, November 14, 2017.
- Jean Twenge. "Changes in how we're spending our free time might explain the unhappiness epidemic" *PsyPost.org*. https://www.psypost.org/2018/08/changes-in-how-were-spending-our-free-time-might-explain-the-unhappiness-epidemic-51893, August 5, 2018.

131

- Table showing birth-year range for seven generations:

Birth Date	Generation Label
1996 to 2015	**Centennials, iGeneration (iGen) or Generation Z (GenZ)**
1977 to 1995	**Millennials or Generation Y (GenY)**
1965 to 1976	**Generation X (GenX)**
1946 to 1964	**Baby Boomers**
1926 to 1945	**Traditionalists or Silent Generation**
1901 to 1925	**G.I. Generation or World War II Generation**
1883 to 1900	**Lost Generation, Generation of 1914 or World War I Generation**

132

- Christine Carter. “Three Risks of Too Much Screen Time for Teens” *Greater Good.* Greater Good Science Center (GGSC), University of California, Berkeley. https://greatergood.berkeley.edu/article/item/three_risks_of_too_much_screen_time_for_teens, November 27, 2018.

133

- Christine Carter. “Three Risks of Too Much Screen Time for Teens” *Greater Good.* Greater Good Science Center (GGSC), University of California, Berkeley. https://greatergood.berkeley.edu/article/item/three_risks_of_too_much_screen_time_for_teens, November 27, 2018.

134

- Christine Carter. “Three Risks of Too Much Screen Time for Teens” *Greater Good.* Greater Good Science Center (GGSC), University of California, Berkeley. https://greatergood.berkeley.edu/article/item/three_risks_of_too_much_screen_time_for_teens, November 27, 2018.

135

- Victoria L. Dunckley, MD. “Electronic Screen Syndrome: An Unrecognized Disorder? Screentime and the rise of mental disorders in children.” *Psychology Today*. https://www.psychologytoday.com/us/blog/mental-wealth/201207/electronic-screen-syndrome-unrecognized-disorder, July 23, 2012.

136

- Sarah D. Sparks. “Students Are Behaving Badly in Class. Excessive Screen Time Might Be to Blame” *Education Week*. https://www.edweek.org/leadership/students-are-behaving-badly-in-class-excessive-screen-time-might-be-to-blame/2022/04, April 12, 2022.

137

- Jean M. Twenge, W. Keith Campbell. “Associations between screen time and lower psychological well-being among children and adolescents: Evidence from a population-based study” *Preventive Medicine Reports,* Volume 12, Pages 271-283. Elsevier B.V. https://doi.org/10.1016/j.pmedr.2018.10.003, October 18, 2018.

138

- Jean M. Twenge. “Have Smartphones Destroyed a Generation?” *The Atlantic.* https://www.theatlantic.com/magazine/archive/2017/09/has-the-smartphone-destroyed-a-generation/534198/, September 2017.

139

- Jeff Haden. *The Motivation Myth.* p. 211. New York: Penguin, 2018.

140

- Christine Carter. “Three Risks of Too Much Screen Time for Teens” *Greater Good.* Greater Good Science Center (GGSC), University of California, Berkeley. https://greatergood.berkeley.edu/article/item/three_risks_of_too_much_screen_time_for_teens, November 27, 2018.
- Louise C. Hawkley, Ph.D. ; John T. Cacioppo, Ph.D. “Loneliness Matters: A Theoretical and Empirical Review of Consequences and Mechanisms” *Annals of Behavioral Medicine*, Volume 40, Issue 2, October 2010, Pages 218–227.

141

- Jean M. Twenge, Thomas E. Joiner, Megan L. Rogers, Gabrielle N. Martin. “Increases in Depressive Symptoms, Suicide-Related Outcomes, and Suicide Rates Among U.S. Adolescents After 2010 and Links to Increased New Media Screen Time” *Clinical Psychological Science.* http://journals.sagepub.com/doi/full/10.1177/2167702617723376, November 14, 2017.

142

- Adam Satariano. “British Ruling Pins Blame on Social Media for Teenager’s Suicide” *The New York Times.* https://www.nytimes.com/2022/10/01/business/instagram-suicide-ruling-britain.html, October 1, 2022.

143

- Ethan Kross, Philippe Verduyn, Emre Demiralp, Jiyoung Park, David Seungjae Lee, Natalie Lin, Holly Shablack, John Jonides, Oscar Ybarra. “Facebook Use Predicts Declines in Subjective Well-Being in Young Adults” *PLoS ONE* 8(8): e69841. doi:10.1371/journal.pone.0069841. https://journals.plos.org/plosone/article?id=10.1371/journal.pone.0069841, August 14, 2013.

144

- Gadi Lissak. “Adverse physiological and psychological effects of screen time on children and adolescents: Literature review and case study” *Environmental Research,* Volume 164, July 2018, Pages 149-157. https://doi.org/10.1016/j.envres.2018.01.015 (also https://www.sciencedirect.com/science/article/pii/S001393511830015X), February, 27 2018.

145

- Melissa G. Hunt, Rachel Marx, Courtney Lipson, Jordyn Young. “No More FOMO: Limiting Social Media Decreases Loneliness and Depression” *Journal of Social and Clinical Psychology*. Guilford Press. https://guilfordjournals.com/doi/10.1521/jscp.2018.37.10.751, October 18, 2018.

146

- Jean M. Twenge, Thomas E. Joiner, Megan L. Rogers, Gabrielle N. Martin. “Increases in Depressive Symptoms, Suicide-Related Outcomes, and Suicide Rates Among U.S. Adolescents After 2010 and Links to Increased New Media Screen Time” *Clinical Psychological Science.* http://journals.sagepub.com/doi/full/10.1177/2167702617723376, November 14, 2017.

147

- Eric Pickersgill. *Removed.* http://www.removed.social/, accessed October 21, 2015.

148

- Richard Freed. "The Tech Industry's War on Kids: How psychology is being used as a weapon against children" *Blog, Child and adolescent psychologist* Richard Freed. https://medium.com/@richardnfreed/the-tech-industrys-psychological-war-on-kids-c452870464ce, March 12, 2018.

149

- Elia Abi-Jaoude, MSc MD; Karline Treurnicht Naylor, MPH MD; and Antonio Pignatiello, MD. "Smartphones, social media use and youth mental health" US National Library of Medicine, National Institutes of Health, *Canadian Medical Association Journal*, Vol. 192(6): E136–E141. https://www.ncbi.nlm.nih.gov/pmc/articles/PMC7012622/, February 10, 2020. https://www.cmaj.ca/content/192/6/E136, February 10, 2020.
- Jean M. Twenge. "Have Smartphones Destroyed a Generation?" *The Atlantic*. https://www.theatlantic.com/magazine/archive/2017/09/has-the-smartphone-destroyed-a-generation/534198/, September 2017.
- Jean M. Twenge, W. Keith Campbell. "Associations between screen time and lower psychological well-being among children and adolescents: Evidence from a population-based study" *Preventive Medicine Reports,* Volume 12, Pages 271-283. Elsevier B.V. https://doi.org/10.1016/j.pmedr.2018.10.003, October 18, 2018.
- Jean M. Twenge, Thomas E. Joiner, Megan L. Rogers, Gabrielle N. Martin. "Increases in Depressive Symptoms, Suicide-Related Outcomes, and Suicide Rates Among U.S. Adolescents After 2010 and Links to Increased New Media Screen Time" *Clinical Psychological Science.* http://journals.sagepub.com/doi/full/10.1177/2167702617723376, November 14, 2017.

- Ethan Kross, Philippe Verduyn, Emre Demiralp, Jiyoung Park, David Seungjae Lee, Natalie Lin, Holly Shablack, John Jonides, Oscar Ybarra. "Facebook Use Predicts Declines in Subjective Well-Being in Young Adults" *PLoS ONE* 8(8): e69841. doi:10.1371/journal.pone.0069841. https://journals.plos.org/plosone/article?id=10.1371/journal.pone.0069841, August 14, 2013.

150

- Hilary Andersson. "Social media apps are 'deliberately' addictive to users" *BBC.* https://www.bbc.com/news/technology-44640959, July 4, 2018.

151

- Adam Alter. "Tech Bigwigs Know How Addictive Their Products Are. Why Don't the Rest of Us?" *Wired.com.* https://www.wired.com/2017/03/irresistible-the-rise-of-addictive-technology-and-the-business-of-keeping-us-hooked/, March 24, 2017.

152

- Elise Hu. "Facebook Makes Us Sadder And Less Satisfied, Study Finds" *National Public Radio.* http://www.npr.org/blogs/alltechconsidered/2013/08/19/213568763/researchers-facebook-makes-us-sadder-and-less-satisfied, August 20, 2013.

153

- Dennis Thompson. "More Americans suffering from stress, anxiety and depression, study finds" *CBS News - Healthday.* http://www.cbsnews.com/news/stress-anxiety-depression-mental-illness-increases-study-finds/, April 17, 2017.
- Mel Schwartz. "Is Our Society Manufacturing Depressed People?" *Psychology Today.* https://www.psychologytoday.com/blog/shift-mind/201203/is-our-society-manufacturing-depressed-people, March 19, 2012.
- Jean M. Twenge, Thomas E. Joiner, Megan L. Rogers, Gabrielle N. Martin. "Increases in Depressive Symptoms, Suicide-Related Outcomes, and Suicide Rates Among U.S. Adolescents After 2010 and Links to Increased New Media Screen Time" *Clinical Psychological Science.* http://journals.sagepub.com/doi/full/10.1177/2167702617723376, November 14, 2017.

154

- Jean M. Twenge, Thomas E. Joiner, Megan L. Rogers, Gabrielle N. Martin. "Increases in Depressive Symptoms, Suicide-Related Outcomes, and Suicide Rates Among U.S. Adolescents After 2010 and Links to Increased New Media Screen Time" *Clinical Psychological Science.* http://journals.sagepub.com/doi/full/10.1177/2167702617723376, November 14, 2017.

155

- Dennis Thompson. "More Americans suffering from stress, anxiety and depression, study finds" *CBS News - Healthday.* http://www.cbsnews.com/news/stress-anxiety-depression-mental-illness-increases-study-finds/, April 17, 2017.

156

- David Brooks. "Why Your Social Life Is Not What It Should Be" *The New York Times.* https://www.nytimes.com/2022/08/25/opinion/social-life-talk-strangers.html, August 25, 2022.

157

- Nicholas Epley and Juliana Schroeder. "Mistakenly Seeking Solitude" *Journal of Experimental Psycholog*y, American Psychological Association. 2014, Vol. 143, No. 5, pp. 1980–1999. https://uploads-ssl.webflow.com/5c484e0f4aa6f839dc553c45/5c9540a7e291048443c0c276_EpleySchroederJEPG2014.pdf https://doi.org/10.1037/a0037323, accessed August 29, 2022.
- David Brooks. "Why Your Social Life Is Not What It Should Be" *The New York Times.* https://www.nytimes.com/2022/08/25/opinion/social-life-talk-strangers.html, August 25, 2022.

158

- Nicholas Epley, Michael Kardas, Xuan Zhao, Stav Atir, Juliana Schroeder. "Undersociality: miscalibrated social cognition can inhibit social connection" *Trends in Cognitive Sciences*. Vol. 26, Issue 5, pp. 406–418, May 1, 2022. https://doi.org/10.1016/j.tics.2022.02.007, March 24, 2022.
- David Brooks. "Why Your Social Life Is Not What It Should Be" *The New York Times.* https://www.nytimes.com/2022/08/25/opinion/social-life-talk-strangers.html, August 25, 2022.

159

- Stav Atir, Kristina A. Wald, and Nicholas Epley. "Talking with strangers is surprisingly informative" *Proceedings of the National Academy of Sciences*. Vol. 119, No. 34. https://www.pnas.org/doi/10.1073/pnas.2206992119, August 16, 2022.
- David Brooks. "Why Your Social Life Is Not What It Should Be" *The New York Times.* https://www.nytimes.com/2022/08/25/opinion/social-life-talk-strangers.html, August 25, 2022.

160

- M. Kardas, J. Schroeder, and E. O'Brien. "Keep talking: (Mis)understanding the hedonic trajectory of conversation" *Journal of Personality and Social Psychology*. Advance online publication. https://psycnet.apa.org/record/2022-16089-001?doi=1 https://psycnet.apa.org/doiLanding?doi=10.1037%2Fpspi0000379, accessed August 29, 2022.
- David Brooks. "Why Your Social Life Is Not What It Should Be" *The New York Times.* https://www.nytimes.com/2022/08/25/opinion/social-life-talk-strangers.html, August 25, 2022.

161

- Michael Kardas, Amit Kumar, and Nicholas Epley. "Overly Shallow?: Miscalibrated Expectations Create a Barrier to Deeper Conversation" *Personality and Social Psychology: Attitudes and Social Cognition*. American Psychological Association. https://www.apa.org/pubs/journals/releases/psp-pspa0000281.pdf, May 4, 2021.
- David Brooks. "Why Your Social Life Is Not What It Should Be" *The New York Times.* https://www.nytimes.com/2022/08/25/opinion/social-life-talk-strangers.html, August 25, 2022.

162

- Travis Bradberry. "13 Habits of Exceptionally Likeable People". https://www.linkedin.com/pulse/13-habits-exceptionally-likeable-people-dr-travis-bradberry, January 27, 2015.

163

- Jean M. Twenge, Thomas E. Joiner, Megan L. Rogers, Gabrielle N. Martin. "Increases in Depressive Symptoms, Suicide-Related Outcomes, and Suicide Rates Among U.S. Adolescents After 2010 and Links to Increased New Media Screen Time" *Clinical Psychological Science.* http://journals.sagepub.com/doi/full/10.1177/2167702617723376, November 14, 2017.

164

- Jamil Zaki. "What, Me Care? Young Are Less Empathetic" *Scientific American*. A recent study finds a decline in empathy among young people in the U.S. http://www.scientificamerican.com/article/what-me-care/, December 23, 2010.

- "Children who watch 'excessive' amounts of TV are more likely to have criminal convictions, exhibit aggression and experience negative emotions: study." *New York Daily News.* http://www.nydailynews.com/life-style/health/kids-watch-excessive-tv-criminal-convictions-young-adulthood-study-article-1.1267868, February 19, 2013.
  "With every hour in front of the television, kids were more likely to show aggressive behavior or receive a criminal conviction by early adulthood, according to a study published in 'Pediatrics.' The issue isn't necessarily the content of the programming, but the social isolation that comes from so many hours in front of the tube."
- "TV retards your child's development" *Consumers Association of Penang.* http://www.consumer.org.my/index.php/development/education/347-tv-retards-your-childs-development, accessed July 17, 2016.
  Also See:
  "More arguments against TV" Consumers Association of Penang. https://consumer.org.my/more-arguments-against-tv/, March 18, 2022.
- Emma Henderson. "Watching lots of TV 'makes you stupid'" *The Independent.* http://www.independent.co.uk/news/science/watching-lots-of-tv-makes-you-stupid-says-american-universities-a6759026.html, December 3, 2015.
- Herbert E. Krugman and Eugene L. Hartley. "Passive Learning From Television" *The Public Opinion Quarterly,* Vol. 34, No. 2. (Oxford University Press, 1970) pp. 184-190.
- Herbert E. Krugman. "Brain Wave Measures of Media Involvement" *How Advertising Works: The Role of Research.* (New York SAGE Publications, 1998) pp. 139–151.
- "Your brain waves change when you watch TV" *I Am Awake.* http://www.iamawake.co/your-brain-waves-change-when-you-watch-tv/, October 11, 2013.
- Kristina Birdsong. "This is Your Child's Brain on TV" *Scientific Learning.* http://www.scilearn.com/blog/how-television-impacts-learning, March 22, 2016.
  http://54.186.226.228/blog/your-childs-brain-tv, March 22, 2016.
- Douglas Fields. "Watching TV Alters Children's Brain Structure and Lowers IQ". http://rdouglasfields.com/2015/05/watching-tv-alters-childrens-brain-structure-and-lowers-iq/, May 4, 2015.

- David L. Hill. “Why to Avoid TV Before Age 2” (early brain development). *American Academy of Pediatrics*. http://www.healthychildren.org/English/family-life/Media/Pages/Why-to-Avoid-TV-Before-Age-2.aspx, May 11, 2013.
  Original Source: Dad to Dad: Parenting Like a Pro (Copyright © American Academy of Pediatrics 2012).
- Alice Park. “Baby Einsteins: Not So Smart After All” *Time*. http://content.time.com/time/health/article/0,8599,1650352,00.html August 06, 2007.
- “Television vs. Reading” *Parent Soup*®, a Trademark of iVillage℠ Inc. Copyright 1996. http://webshare.northseattle.edu/fam180/topics/tv/tvvsread.htm, accessed May 29, 2014.
- Jim Trelease. *The Read-Aloud Handbook*, Penguin Books, 1995.
- Marie Winn. “Television and the Brain” and “Brain Changes” *Plug-In Drug*. (New York: Viking Penguin, 1977, Revised and Updated Edition, 2002) pp. 67–69:
  “‘Research conducted during the next two decades removed any doubt about the impact of early brain stimulation on a child’s later cognitive development. … they were able to demonstrate that environmental factors can alter neuron pathways during early childhood and long after … among the most important of the environmental factors … are the language and eye contact an infant is exposed to [and] … the number of words an infant hears each day is the single most important predictor of later intelligence, school success and social competence. But there’s one catch. The words have to come from an attentive, engaged human being … radio and television do not work.”

165

- Maia Szalavitz. “Misery Has More Company Than You Think, Especially on Facebook” *Time*. http://healthland.time.com/2011/01/27/youre-not-alone-misery-has-more-company-than-you-think/, January 27, 2011.

- Alexander H. Jordan, Benoît Monin, Carol S. Dweck, Benjamin J. Lovett, Oliver P. John, and James J. Gross. “Misery Has More Company Than People Think: Underestimating the Prevalence of Others’ Negative Emotions” *Personality and Social Psychology Bulletin* January 2011 37: 120-135. National Institutes of Health. http://www.ncbi.nlm.nih.gov/pmc/articles/PMC4138214/, January 2011, latest update August 19, 2014.
- Elise Hu. “Facebook Makes Us Sadder And Less Satisfied, Study Finds” *National Public Radio*. http://www.npr.org/blogs/alltechconsidered/2013/08/19/213568763/researchers-facebook-makes-us-sadder-and-less-satisfied, August 20, 2013. “If you’re feeling bummed, researchers did test for and find a solution. The prescription for Facebook despair is less Facebook. Researchers found that face-to-face or phone interaction — those outmoded, analog ways of communication — had the opposite effect. Direct interactions with other human beings led people to feel better.”
- Ethan Kross, Philippe Verduyn, Emre Demiralp, Jiyoung Park, David Seungjae Lee, Natalie Lin, Holly Shablack, John Jonides, Oscar Ybarra. “Facebook Use Predicts Declines in Subjective Well-Being in Young Adults” *PLoS ONE* 8(8): e69841. doi:10.1371/journal.pone.0069841. http://www.plosone.org/article/info%3Adoi%2F10.1371%2Fjournal.pone.0069841, August 14, 2013.

166

- Herbert E. Krugman and Eugene L. Hartley. “Passive Learning From Television” *The Public Opinion Quarterly,* Vol. 34, No. 2. (Oxford University Press, 1970) pp. 184-190.
- Herbert E. Krugman. “Brain Wave Measures of Media Involvement” *How Advertising Works: The Role of Research.* (New York SAGE Publications, 1998) pp. 139–151.
- “Your brain waves change when you watch TV” *I Am Awake.* http://www.iamawake.co/your-brain-waves-change-when-you-watch-tv/, October 11, 2013.

167

- Jean M. Twenge, Thomas E. Joiner, Megan L. Rogers, Gabrielle N. Martin. "Increases in Depressive Symptoms, Suicide-Related Outcomes, and Suicide Rates Among U.S. Adolescents After 2010 and Links to Increased New Media Screen Time" *Clinical Psychological Science.* http://journals.sagepub.com/doi/full/10.1177/2167702617723376, November 14, 2017.

168

- Robert Kubey and Mihaly Csikszentmihalyi. "Television Addiction is no mere metaphor" *Scientific American.* http://www.academia.edu/5065840/Television_Addiction_is_no_mere_metaphor, accessed January 23, 2016.

169

- Nicholas Kardaras. "It's 'digital heroin': How screens turn kids into psychotic junkies" *The New York Post.* https://nypost.com/2016/08/27/its-digital-heroin-how-screens-turn-kids-into-psychotic-junkies/, August 27, 2016.

170

- Marie Winn. "Losing the Thread," "The Basic Building Blocks," "A Preference for Watching" *Plug-In Drug.* (New York: Viking Penguin, 1977, Revised and Updated Edition, 2002) pp. 95–99.

171

- Robert Kubey and Mihaly Csikszentmihalyi. "Television Addiction is no mere metaphor" *Scientific American.* http://www.academia.edu/5065840/Television_Addiction_is_no_mere_metaphor, accessed January 23, 2016.

172

- Robert Kubey and Mihaly Csikszentmihalyi. "Television Addiction is no mere metaphor" *Scientific American.* http://www.academia.edu/5065840/Television_Addiction_is_no_mere_metaphor, accessed January 23, 2016.

173

- Jessica Stillman. "The Cold, Hard Truth: You're Overwhelmed Because You Want to Be" *Inc.com.* http://www.inc.com/jessica-stillman/cure-for-feeling-overwhelmed-is-to-stop-talking-about-it.html, March 28, 2014.

174

- Jessica Stillman. "The Cold, Hard Truth: You're Overwhelmed Because You Want to Be" *Inc.com.* http://www.inc.com/jessica-stillman/cure-for-feeling-overwhelmed-is-to-stop-talking-about-it.html, March 28, 2014.

175

- Jessica Stillman. "Complaining Is Terrible for You, According to Science". *Inc.com*. http://www.inc.com/jessica-stillman/complaining-rewires-your-brain-for-negativity-science-says.html, February 29, 2016.

176

- William F. Doverspike, Ph.D. "How To Make Yourself Miserable: Discovering the Secrets to Unhappiness" *Georgia Psychological Association*. Atlanta. http://gapsychology.org/displaycommon.cfm?an=1&subarticlenbr=341, accessed September 29, 2014.
  Also: Drdoverspike.com. http://drwilliamdoverspike.com/files/how_to_make_yourself_miserable_revised_version.pdf.
- Belinda Goldsmith. "Watching hours of TV daily could shorten your life – study" Ed. Miral Fahmy. *Reuters*. http://in.reuters.com/article/worldNews/idINIndia-45316820100111, January 12, 2010.

177

- Travis Bradberry. "13 Things Mentally Strong People Won't Do". https://www.linkedin.com/pulse/13-things-mentally-strong-people-wont-do-dr-travis-bradberry/, September 11, 2017.

178

- William F. Doverspike, Ph.D. "How To Make Yourself Miserable: Discovering the Secrets to Unhappiness" *Georgia Psychological Association*. Atlanta. http://gapsychology.org/displaycommon.cfm?an=1&subarticlenbr=341, accessed September 29, 2014.
  Also: Drdoverspike.com. http://drwilliamdoverspike.com/files/how_to_make_yourself_miserable_revised_version.pdf.

179

- "TV retards your child's development" *Consumers Association of Penang.* http://www.consumer.org.my/index.php/development/education/347-tv-retards-your-childs-development, accessed July 17, 2016.
  Also See:
  "More arguments against TV" Consumers Association of Penang. https://consumer.org.my/more-arguments-against-tv/, March 18, 2022.
- Emma Henderson. "Watching lots of TV 'makes you stupid'" *The Independent.* http://www.independent.co.uk/news/science/watching-lots-of-tv-makes-you-stupid-says-american-universities-a6759026.html, December 3, 2015.
- Herbert E. Krugman and Eugene L. Hartley. "Passive Learning From Television" *The Public Opinion Quarterly,* Vol. 34, No. 2. (Oxford University Press, 1970) pp. 184-190.
- Herbert E. Krugman. "Brain Wave Measures of Media Involvement" *How Advertising Works: The Role of Research.* (New York SAGE Publications, 1998) pp. 139–151.
- "Your brain waves change when you watch TV" *I Am Awake.* http://www.iamawake.co/your-brain-waves-change-when-you-watch-tv/, October 11, 2013.
- Kristina Birdsong. "This is Your Child's Brain on TV" *Scientific Learning.* http://www.scilearn.com/blog/how-television-impacts-learning, March 22, 2016.
  http://54.186.226.228/blog/your-childs-brain-tv, March 22, 2016.
- Douglas Fields. "Watching TV Alters Children's Brain Structure and Lowers IQ". http://rdouglasfields.com/2015/05/watching-tv-alters-childrens-brain-structure-and-lowers-iq/, May 4, 2015.

- Jamil Zaki. "What, Me Care? Young Are Less Empathetic" *Scientific American*. A recent study finds a decline in empathy among young people in the U.S. http://www.scientificamerican.com/article/what-me-care/, December 23, 2010.
- "Children who watch 'excessive' amounts of TV are more likely to have criminal convictions, exhibit aggression and experience negative emotions: study." *New York Daily News*. http://www.nydailynews.com/life-style/health/kids-watch-excessive-tv-criminal-convictions-young-adulthood-study-article-1.1267868, February 19, 2013.
  "With every hour in front of the television, kids were more likely to show aggressive behavior or receive a criminal conviction by early adulthood, according to a study published in 'Pediatrics.' The issue isn't necessarily the content of the programming, but the social isolation that comes from so many hours in front of the tube."
- David L. Hill. "Why to Avoid TV Before Age 2" (early brain development). *American Academy of Pediatrics*. http://www.healthychildren.org/English/family-life/Media/Pages/Why-to-Avoid-TV-Before-Age-2.aspx, May 11, 2013.
  Original Source: Dad to Dad: Parenting Like a Pro (Copyright © American Academy of Pediatrics 2012).
- Alice Park. "Baby Einsteins: Not So Smart After All" *Time*. http://content.time.com/time/health/article/0,8599,1650352,00.html August 06, 2007.
- "Television vs. Reading" *Parent Soup*®, a Trademark of iVillage℠ Inc. Copyright 1996. http://webshare.northseattle.edu/fam180/topics/tv/tvvsread.htm, accessed May 29, 2014.
- Jim Trelease. *The Read-Aloud Handbook*, Penguin Books, 1995.

- Edward L. Schor. *Caring for your school-age child : ages 5 to 12.* American Academy of Pediatrics. New York : Bantam Books, 2004.
  "What Children are NOT Doing When Watching TV" American Academy of Pediatrics.
  http://www.healthychildren.org/English/family-life/Media/Pages/What-Children-are-NOT-Doing-When-Watching-TV.aspx, May 11, 2013.
  Also See: *Caring for Our Children*. A Joint Collaborative Project of the American Academy of Pediatrics and the American Public Health Association.
  https://nrckids.org/files/CFOC4%20pdf-%20FINAL.pdf, 2019.
- Tina D. Hoang, MSPH; Jared Reis, PhD; Na Zhu, MD, MPH; David R. Jacobs Jr, PhD; Lenore J. Launer, PhD; Rachel A. Whitmer, PhD; Stephen Sidney, MD; Kristine Yaffe, MD.
  "Effect of Early Adult Patterns of Physical Activity and Television Viewing on Midlife Cognitive Function" *JAMA Psychiatry* (The Journal of the American Medical Association).
  http://archpsyc.jamanetwork.com/article.aspx?articleid=2471270, January 2016, Vol 73, No. 1.
- Marie Winn. "Television and the Brain," "Brain Changes," "Losing the Thread," "The Basic Building Blocks," "A Preference for Watching," "Free Time and Resourcefulness" *Plug-In Drug.* (New York: Viking Penguin, 1977, Revised and Updated Edition, 2002) pp. 67–69, 95–99, 131.

180

- Claudine Ryan. "Why is watching TV so bad for you?" *ABC Health & Wellbeing*.
  http://www.abc.net.au/health/thepulse/stories/2014/07/24/4043618.htm July 24, 2014.

181

- Gadi Lissak. "Adverse physiological and psychological effects of screen time on children and adolescents: Literature review and case study" *Environmental Research,* Volume 164, July 2018, Pages 149-157.
  https://doi.org/10.1016/j.envres.2018.01.015
  (also
  https://www.sciencedirect.com/science/article/pii/S001393511830015X), February, 27 2018.

182

- Amanda Gardner. "TV watching raises risk of health problems, dying young" *CNN*. http://www.cnn.com/2011/HEALTH/06/14/tv.watching.unhealthy/, June 14, 2011.
    - "For every two hours Americans spend watching TV each day, there are 176 new cases of diabetes, 38 additional deaths from heart disease, and 104 additional deaths due to any cause per 100,000 people per year"—That's 2 hours/day per 100,000 people. That means given the US population of 324,000,000, and the fact that average US TV viewing is 4 hours per day—TV viewing causes a total of 1,140,480 new cases of diabetes, 246,240 additional deaths from heart disease, and 673,920 additional deaths, every year.

183

- Amanda Gardner. "TV watching raises risk of health problems, dying young" *CNN*. http://www.cnn.com/2011/HEALTH/06/14/tv.watching.unhealthy/, June 14, 2011.
    - "For every two hours Americans spend watching TV each day, there are 176 new cases of diabetes, 38 additional deaths from heart disease, and 104 additional deaths due to any cause per 100,000 people per year"—That's 2 hours/day per 100,000 people. That means given the US population of 324,000,000, and the fact that average US TV viewing is 4 hours per day—TV viewing causes a total of 1,140,480 new cases of diabetes, 246,240 additional deaths from heart disease, and 673,920 additional deaths, every year.

184

- Jason M. Nagata MD, MSc; Puja Iyer BA; Jonathan Chu BA; Fiona C. Baker PhD; Kelley Pettee Gabriel MS, PhD; Andrea K. Garber PhD, RD; Stuart B. Murray DClinPsych, PhD; Kirsten Bibbins-Domingo PhD, MD, MAS; Kyle T. Ganson PhD, MSW. "Contemporary screen time modalities among children 9–10 years old and binge-eating disorder at one-year follow-up: A prospective cohort study" *International Journal of Eating Disorders*, Vol. 54, Issue 5, pages 887–892. https://onlinelibrary.wiley.com/doi/10.1002/eat.23489, March 1, 2021.

185

- Jason M. Nagata; Puja Iyer; Jonathan Chu; Fiona C. Baker; Kelley Pettee Gabriel; Andrea K. Garber; Stuart B. Murray; Kirsten Bibbins-Domingo; Kyle T. Ganson. "Contemporary screen time usage among children 9–10-years-old is associated with higher body mass index percentile at 1-year follow-up: A prospective cohort study" *Pediatric Obesity*, Vol. 16, Issue 12. https://onlinelibrary.wiley.com/doi/10.1111/ijpo.12827, June 28, 2021.

186

- Allan Stromfeldt Christensen. "Lemminged: to be herded off the peak oil cliff by filmmakers" *TransitionVoice.com.* http://transitionvoice.com/2014/11/lemminged-to-be-herded-off-the-peak-oil-cliff-by-filmmakers/, November 11, 2014.

187

- "Consumer Culture" *Wikipedia.* https://en.wikipedia.org/wiki/Consumer_Culture, accessed December 15, 2018.

188

- "The Rise of American Consumerism" *PBS.ORG.* https://www.pbs.org/wgbh/americanexperience/features/tupperware-consumer/, accessed December 15, 2018.

189

- "Television History - The First 75 Years" *TVhistory.TV.* © 2001-2013. http://www.tvhistory.tv/facts-stats.htm.

190

- Allan Stromfeldt Christensen. "Lemminged: to be herded off the peak oil cliff by filmmakers" *TransitionVoice.com.* http://transitionvoice.com/2014/11/lemminged-to-be-herded-off-the-peak-oil-cliff-by-filmmakers/, November 11, 2014.

191

- Emily Guy Birken. "This Is How Americans Spent Their Money in the 1950s" *WiseBread.* https://www.wisebread.com/this-is-how-americans-spent-their-money-in-the-1950s, January 12, 2016.

192

- Emily Guy Birken. "This Is How Americans Spent Their Money in the 1950s" *WiseBread.* https://www.wisebread.com/this-is-how-americans-spent-their-money-in-the-1950s, January 12, 2016.

193

- Chris Kirk. "Five Charts That Show Americans Families' Debt Crisis" *Slate.com.* (Data from the Federal Reserve Bank, St. Louis.) http://www.slate.com/articles/business/the_united_states_of_debt/2016/05/the_rise_of_household_debt_in_the_u_s_in_five_charts.html, May 12, 2016.

194

- Allan Stromfeldt Christensen. "Lemminged: to be herded off the peak oil cliff by filmmakers" *TransitionVoice.com.* http://transitionvoice.com/2014/11/lemminged-to-be-herded-off-the-peak-oil-cliff-by-filmmakers/, November 11, 2014.

195

- Allan Stromfeldt Christensen. "Lemminged: to be herded off the peak oil cliff by filmmakers" *TransitionVoice.com.* http://transitionvoice.com/2014/11/lemminged-to-be-herded-off-the-peak-oil-cliff-by-filmmakers/, November 11, 2014.

196

- Allan Stromfeldt Christensen. "Lemminged: to be herded off the peak oil cliff by filmmakers" *TransitionVoice.com.* http://transitionvoice.com/2014/11/lemminged-to-be-herded-off-the-peak-oil-cliff-by-filmmakers/, November 11, 2014.

197

- Allan Stromfeldt Christensen. "Lemminged: to be herded off the peak oil cliff by filmmakers" *TransitionVoice.com.* http://transitionvoice.com/2014/11/lemminged-to-be-herded-off-the-peak-oil-cliff-by-filmmakers/, November 11, 2014.

- Eric Holthaus and Chris Kirk. “A Filthy History: Interactive map: Which countries have emitted the most carbon since 1850?” *Slate.com*. http://www.slate.com/articles/technology/future_tense/2014/05/carbon_dioxide_emissions_by_country_over_time_the_worst_global_warming_polluters.html.

198

- Eric Holthaus and Chris Kirk. “A Filthy History: Interactive map: Which countries have emitted the most carbon since 1850?” *Slate.com*. http://www.slate.com/articles/technology/future_tense/2014/05/carbon_dioxide_emissions_by_country_over_time_the_worst_global_warming_polluters.html.

199

- Ron Marshall. “How Many Ads Do You See in One Day?” *Red Crow Marketing.* https://www.redcrowmarketing.com/2015/09/10/many-ads-see-one-day/, September 10, 2015.

200

- Allan Stromfeldt Christensen. “Lemminged: to be herded off the peak oil cliff by filmmakers” *TransitionVoice.com.* http://transitionvoice.com/2014/11/lemminged-to-be-herded-off-the-peak-oil-cliff-by-filmmakers/, November 11, 2014.
- Eric Holthaus and Chris Kirk. “A Filthy History: Interactive map: Which countries have emitted the most carbon since 1850?” *Slate.com*. http://www.slate.com/articles/technology/future_tense/2014/05/carbon_dioxide_emissions_by_country_over_time_the_worst_global_warming_polluters.html.

201

- Eric Holthaus and Chris Kirk. “A Filthy History: Interactive map: Which countries have emitted the most carbon since 1850?” *Slate.com*. http://www.slate.com/articles/technology/future_tense/2014/05/carbon_dioxide_emissions_by_country_over_time_the_worst_global_warming_polluters.html.

202

- "Carbon footprint" *Wikipedia*. http://en.wikipedia.org/wiki/Carbon_footprint, accessed November 1, 2014.

203

- Eric Holthaus and Chris Kirk. "A Filthy History: Interactive map: Which countries have emitted the most carbon since 1850?" *Slate.com*. http://www.slate.com/articles/technology/future_tense/2014/05/carbon_dioxide_emissions_by_country_over_time_the_worst_global_warming_polluters.html.

204

- Eric A. Taub. "E.V.s Start With a Bigger Carbon Footprint. But That Doesn't Last" *The New York Times*. https://www.nytimes.com/2022/10/19/business/electric-vehicles-carbon-footprint-batteries.html, November 7, 2022.
- Maxwell Woody et al. "Corrigendum: The role of pickup truck electrification in the decarbonization of light-duty vehicles" *Environmental Research Letters* 17 089501. *IOP Publishing Ltd*. https://iopscience.iop.org/article/10.1088/1748-9326/ac7cfc, July 15, 2022.

205

- "How Satellites Work With Mobile Phones" *CompareMyMobile*. http://blog.comparemymobile.com/how-satellites-work-with-mobile-phones/, accessed December 21, 2017.

206

- Adeel Younas. "Report: Smartphones Are Destroying Our Planet Faster Than We Think" *TechWafer*. https://techwafer.com/report-smartphones-are-destroying-our-planet-faster-than-we-think/, April 2, 2018.
- Lotfi Belkhir and Ahmed Elmeligi. "Assessing ICT global emissions footprint: Trends to 2040 & recommendations" *Journal of Cleaner Production*, Volume 177, Pages 448-463. Elsevier B.V. https://doi.org/10.1016/j.jclepro.2017.12.239, March 10, 2018.

- James Suckling and Jacquetta Lee. "Redefining scope: the true environmental impact of smartphones?" *The International Journal of Life Cycle Assessment*, Volume 20, Issue 8, pp 1181–1196. Springer Berlin Heidelberg.
  https://doi.org/10.1016/j.jclepro.2017.12.239, June 10, 2015.
  Also Updated:
  Belkhir, Lotfi, and Ahmed Elmeligi. "Assessing ICT global emissions footprint: Trends to 2040 & recommendations" *Journal of Cleaner Production*. Volume 177, pp 448-463.
  https://www.sciencedirect.com/science/article/abs/pii/S095965261733233X, Online January 2, 2018; Version of Record January 3, 2018.

207

- Christopher Alexander; Sara Ishikawa; Murray Silverstein. *A Pattern Language* (p. 618), Oxford University Press, 1977.

208

- Victoria L. Dunckley, MD. "Screentime and Arrested Social Development" *Psychology Today.*
  https://www.psychologytoday.com/us/blog/mental-wealth/201606/screentime-and-arrested-social-development, June 30, 2016.
- Rosalina Richards, Rob McGee, Sheila M Williams, David Welch, and Robert J Hancox. "Adolescent Screen Time and Attachment to Parents and Peers" *Archives of Pediatrics & Adolescent Medicine.* 164, no. 3 (March 2010): 258–62. doi:10.1001/archpediatrics.2009.280.
  https://jamanetwork.com/journals/jamapediatrics/fullarticle/382905, March 1, 2010.

209

- Victoria L. Dunckley, MD. "Screentime and Arrested Social Development" *Psychology Today.*
  https://www.psychologytoday.com/us/blog/mental-wealth/201606/screentime-and-arrested-social-development, June 30, 2016.

210

- Victoria L. Dunckley, MD. "Screentime and Arrested Social Development" *Psychology Today.* https://www.psychologytoday.com/us/blog/mental-wealth/201606/screentime-and-arrested-social-development, June 30, 2016.

211

- Azmaira H. Maker. "Screen Time: The Impact on Kids and Parenting: New research explains the significant negative effects of excessive screen time." *Psychology Today.* https://www.psychologytoday.com/us/blog/helping-kids-cope/201808/screen-time-the-impact-kids-and-parenting, August 19, 2018.

212

- Azmaira H. Maker. "Screen Time: The Impact on Kids and Parenting: New research explains the significant negative effects of excessive screen time." *Psychology Today.* https://www.psychologytoday.com/us/blog/helping-kids-cope/201808/screen-time-the-impact-kids-and-parenting, August 19, 2018.

213

- "Children who watch 'excessive' amounts of TV are more likely to have criminal convictions, exhibit aggression and experience negative emotions: study." *New York Daily News.* http://www.nydailynews.com/life-style/health/kids-watch-excessive-tv-criminal-convictions-young-adulthood-study-article-1.1267868, February 19, 2013.
  "With every hour in front of the television, kids were more likely to show aggressive behavior or receive a criminal conviction by early adulthood, according to a study published in 'Pediatrics.' The issue isn't necessarily the content of the programming, but the social isolation that comes from so many hours in front of the tube."

214

- Victoria L. Dunckley, MD. "Screentime and Arrested Social Development" *Psychology Today.* https://www.psychologytoday.com/us/blog/mental-wealth/201606/screentime-and-arrested-social-development, June 30, 2016.

215

- "TV retards your child's development" *Consumers Association of Penang.* http://www.consumer.org.my/index.php/development/education/347-tv-retards-your-childs-development, accessed July 17, 2016.
  Also See:
  "More arguments against TV" Consumers Association of Penang. https://consumer.org.my/more-arguments-against-tv/, March 18, 2022.
- Emma Henderson. "Watching lots of TV 'makes you stupid'" *The Independent.* http://www.independent.co.uk/news/science/watching-lots-of-tv-makes-you-stupid-says-american-universities-a6759026.html, December 3, 2015.
- Herbert E. Krugman and Eugene L. Hartley. "Passive Learning From Television" *The Public Opinion Quarterly,* Vol. 34, No. 2. (Oxford University Press, 1970) pp. 184-190.
- Herbert E. Krugman. "Brain Wave Measures of Media Involvement" *How Advertising Works: The Role of Research.* (New York SAGE Publications, 1998) pp. 139–151.
- "Your brain waves change when you watch TV" *I Am Awake.* http://www.iamawake.co/your-brain-waves-change-when-you-watch-tv/, October 11, 2013.
- Kristina Birdsong. "This is Your Child's Brain on TV" *Scientific Learning.* http://www.scilearn.com/blog/how-television-impacts-learning, March 22, 2016.
  http://54.186.226.228/blog/your-childs-brain-tv, March 22, 2016.
- Douglas Fields. "Watching TV Alters Children's Brain Structure and Lowers IQ". http://rdouglasfields.com/2015/05/watching-tv-alters-childrens-brain-structure-and-lowers-iq/, May 4, 2015.

- Victoria L. Dunckley, MD. "Gray Matters: Too Much Screen Time Damages the Brain" *Psychology Today.* https://www.psychologytoday.com/us/blog/mental-wealth/201402/gray-matters-too-much-screen-time-damages-the-brain, February 27, 2014.
- Jamil Zaki. "What, Me Care? Young Are Less Empathetic" *Scientific American*. A recent study finds a decline in empathy among young people in the U.S. http://www.scientificamerican.com/article/what-me-care/, December 23, 2010.
- "Children who watch 'excessive' amounts of TV are more likely to have criminal convictions, exhibit aggression and experience negative emotions: study." *New York Daily News.* http://www.nydailynews.com/life-style/health/kids-watch-excessive-tv-criminal-convictions-young-adulthood-study-article-1.1267868, February 19, 2013.
  "With every hour in front of the television, kids were more likely to show aggressive behavior or receive a criminal conviction by early adulthood, according to a study published in 'Pediatrics.' The issue isn't necessarily the content of the programming, but the social isolation that comes from so many hours in front of the tube."
- David L. Hill. "Why to Avoid TV Before Age 2" (early brain development). *American Academy of Pediatrics.* http://www.healthychildren.org/English/family-life/Media/Pages/Why-to-Avoid-TV-Before-Age-2.aspx, May 11, 2013.
  Original Source: Dad to Dad: Parenting Like a Pro (Copyright © American Academy of Pediatrics 2012).
- Alice Park. "Baby Einsteins: Not So Smart After All" *Time.* http://content.time.com/time/health/article/0,8599,1650352,00.html August 06, 2007.
- "Television vs. Reading" *Parent Soup*®, a Trademark of iVillage℠ Inc. Copyright 1996. http://webshare.northseattle.edu/fam180/topics/tv/tvvsread.htm, accessed May 29, 2014.
- Jim Trelease. *The Read-Aloud Handbook*, Penguin Books, 1995.

- Edward L. Schor. *Caring for your school-age child : ages 5 to 12.* American Academy of Pediatrics. New York : Bantam Books, 2004.
  "What Children are NOT Doing When Watching TV" American Academy of Pediatrics.
  http://www.healthychildren.org/English/family-life/Media/Pages/What-Children-are-NOT-Doing-When-Watching-TV.aspx, May 11, 2013.
  Also See: *Caring for Our Children.* A Joint Collaborative Project of the American Academy of Pediatrics and the American Public Health Association.
  https://nrckids.org/files/CFOC4%20pdf-%20FINAL.pdf, 2019.
- Marie Winn. "Television and the Brain," "Brain Changes," "Losing the Thread," "The Basic Building Blocks," "A Preference for Watching," "Free Time and Resourcefulness" *Plug-In Drug.* (New York: Viking Penguin, 1977, Revised and Updated Edition, 2002) pp. 67–69, 95–99, 131. From pp. 67–69:
  "'Research conducted during the next two decades removed any doubt about the impact of early brain stimulation on a child's later cognitive development. ... they were able to demonstrate that environmental factors can alter neuron pathways during early childhood and long after ... among the most important of the environmental factors ... are the language and eye contact an infant is exposed to [and] ... the number of words an infant hears each day is the single most important predictor of later intelligence, school success and social competence. But there's one catch. The words have to come from an attentive, engaged human being ... radio and television do not work."

216

- Azmaira H. Maker. "Screen Time: The Impact on Kids and Parenting: New research explains the significant negative effects of excessive screen time." *Psychology Today.*
  https://www.psychologytoday.com/us/blog/helping-kids-cope/201808/screen-time-the-impact-kids-and-parenting, August 19, 2018.

- Edward L. Schor. *Caring for your school-age child : ages 5 to 12.* American Academy of Pediatrics. New York : Bantam Books, 2004.
  "What Children are NOT Doing When Watching TV" American Academy of Pediatrics.
  http://www.healthychildren.org/English/family-life/Media/Pages/What-Children-are-NOT-Doing-When-Watching-TV.aspx, May 11, 2013.
  Also See: *Caring for Our Children*. A Joint Collaborative Project of the American Academy of Pediatrics and the American Public Health Association.
  https://nrckids.org/files/CFOC4%20pdf-%20FINAL.pdf, 2019.
- Marie Winn. "Losing the Thread," "The Basic Building Blocks," "A Preference for Watching," "Free Time and Resourcefulness" *Plug-In Drug.* (New York: Viking Penguin, 1977, Revised and Updated Edition, 2002) pp. 95–99, 131.
- Alice Park. "Baby Einsteins: Not So Smart After All" *Time*. http://content.time.com/time/health/article/0,8599,1650352,00.html August 06, 2007.
- Rachael Pells. "Today's four-year-olds often 'not physically ready' for school, experts warn" *Independent*. https://www.independent.co.uk/news/education/education-news/school-age-four-year-old-children-not-physically-ready-experts-warn-a7220476.html, September 1, 2016.
- Alexandra Thompson. "Surgery students are struggling to use their hands because they spent 'too much time watching TV' rather than sewing or playing an instrument as children" *The Daily Mail*. https://www.dailymail.co.uk/health/article-6332847/Surgery-students-struggle-use-hands-spent-time-watching-TV.html, October 30, 2018.
- Sukhpreet K. Tamana, Victor Ezeugwu, Joyce Chikuma, Diana L. Lefebvre, Meghan B. Azad, Theo J. Moraes, Padmaja Subbarao, Allan B. Becker, Stuart E. Turvey, Malcolm R. Sears, Bruce D. Dick, Valerie Carson, Carmen Rasmussen, Piush J. Mandhane. "Screen-time is associated with inattention problems in preschoolers: Results from the CHILD birth cohort study" *PLOS ONE Journal of Scientific Research*. https://journals.plos.org/plosone/article?id=10.1371/journal.pone.0213995, April 17, 2019.

- Sheri Madigan, PhD; Dillon Browne, PhD; Nicole Racine, PhD. "Association Between Screen Time and Children's Performance on a Developmental Screening Test" *American Medical Association: JAMA Pediatr. 2019;173(3):244-250. doi:10.1001/jamapediatrics.2018.5056.* https://jamanetwork.com/journals/jamapediatrics/fullarticle/2722666, January 28, 2019.

217

- "Television History - The First 75 Years" *TVhistory.TV.* © 2001-2013. http://www.tvhistory.tv/facts-stats.htm.

218

- "Children who watch 'excessive' amounts of TV are more likely to have criminal convictions, exhibit aggression and experience negative emotions: study" *New York Daily News.* http://www.nydailynews.com/life-style/health/kids-watch-excessive-tv-criminal-convictions-young-adulthood-study-article-1.1267868, February 19, 2013.
- Douglas Fields. "Watching TV Alters Children's Brain Structure and Lowers IQ". http://rdouglasfields.com/2015/05/watching-tv-alters-childrens-brain-structure-and-lowers-iq/, May 4, 2015.

219

- "Children who watch 'excessive' amounts of TV are more likely to have criminal convictions, exhibit aggression and experience negative emotions: study" *New York Daily News.* http://www.nydailynews.com/life-style/health/kids-watch-excessive-tv-criminal-convictions-young-adulthood-study-article-1.1267868, February 19, 2013.
- Douglas Fields. "Watching TV Alters Children's Brain Structure and Lowers IQ". http://rdouglasfields.com/2015/05/watching-tv-alters-childrens-brain-structure-and-lowers-iq/, May 4, 2015.

220

- Jamil Zaki. "What, Me Care? Young Are Less Empathetic" *Scientific American*. A recent study finds a decline in empathy among young people in the U.S. http://www.scientificamerican.com/article/what-me-care/, December 23, 2010.
- "Children who watch 'excessive' amounts of TV are more likely to have criminal convictions, exhibit aggression and experience negative emotions: study." *New York Daily News*. http://www.nydailynews.com/life-style/health/kids-watch-excessive-tv-criminal-convictions-young-adulthood-study-article-1.1267868, February 19, 2013.
  "With every hour in front of the television, kids were more likely to show aggressive behavior or receive a criminal conviction by early adulthood, according to a study published in 'Pediatrics.' The issue isn't necessarily the content of the programming, but the social isolation that comes from so many hours in front of the tube."
- "TV retards your child's development" *Consumers Association of Penang.* http://www.consumer.org.my/index.php/development/education/347-tv-retards-your-childs-development, accessed July 17, 2016.
  Also See:
  "More arguments against TV" Consumers Association of Penang. https://consumer.org.my/more-arguments-against-tv/, March 18, 2022.

221

- Jean M. Twenge. "Have Smartphones Destroyed a Generation?" *The Atlantic*. https://www.theatlantic.com/magazine/archive/2017/09/has-the-smartphone-destroyed-a-generation/534198/, September 2017.

222

- Jamil Zaki. "What, Me Care? Young Are Less Empathetic" *Scientific American*. A recent study finds a decline in empathy among young people in the U.S. http://www.scientificamerican.com/article/what-me-care/, December 23, 2010.

- “Children who watch ‘excessive’ amounts of TV are more likely to have criminal convictions, exhibit aggression and experience negative emotions: study.” *New York Daily News*. http://www.nydailynews.com/life-style/health/kids-watch-excessive-tv-criminal-convictions-young-adulthood-study-article-1.1267868, February 19, 2013.
  “With every hour in front of the television, kids were more likely to show aggressive behavior or receive a criminal conviction by early adulthood, according to a study published in ‘Pediatrics.’ The issue isn’t necessarily the content of the programming, but the social isolation that comes from so many hours in front of the tube.”
- “TV retards your child’s development” *Consumers Association of Penang.* http://www.consumer.org.my/index.php/development/education/347-tv-retards-your-childs-development, accessed July 17, 2016.
  Also See:
  “More arguments against TV” Consumers Association of Penang. https://consumer.org.my/more-arguments-against-tv/, March 18, 2022.

223

- Travis Bradberry. “Eight Habits of Considerate People”. https://www.linkedin.com/pulse/eight-habits-considerate-people-dr-travis-bradberry, November 8, 2017.

224

- Travis Bradberry. “12 Habits of Genuine People”. https://www.linkedin.com/pulse/importance-being-genuine-dr-travis-bradberry/, November 15, 2015.

225

- Jamil Zaki. “What, Me Care? Young Are Less Empathetic” *Scientific American*. A recent study finds a decline in empathy among young people in the U.S. http://www.scientificamerican.com/article/what-me-care/, December 23, 2010.

- "Children who watch 'excessive' amounts of TV are more likely to have criminal convictions, exhibit aggression and experience negative emotions: study." *New York Daily News*. http://www.nydailynews.com/life-style/health/kids-watch-excessive-tv-criminal-convictions-young-adulthood-study-article-1.1267868, February 19, 2013.
  "With every hour in front of the television, kids were more likely to show aggressive behavior or receive a criminal conviction by early adulthood, according to a study published in 'Pediatrics.' The issue isn't necessarily the content of the programming, but the social isolation that comes from so many hours in front of the tube."

226

- Jamil Zaki. "What, Me Care? Young Are Less Empathetic" *Scientific American*. A recent study finds a decline in empathy among young people in the U.S. http://www.scientificamerican.com/article/what-me-care/, December 23, 2010.
- "Children who watch 'excessive' amounts of TV are more likely to have criminal convictions, exhibit aggression and experience negative emotions: study." *New York Daily News*. http://www.nydailynews.com/life-style/health/kids-watch-excessive-tv-criminal-convictions-young-adulthood-study-article-1.1267868, February 19, 2013.
  "With every hour in front of the television, kids were more likely to show aggressive behavior or receive a criminal conviction by early adulthood, according to a study published in 'Pediatrics.' The issue isn't necessarily the content of the programming, but the social isolation that comes from so many hours in front of the tube."
- "TV retards your child's development" *Consumers Association of Penang.* http://www.consumer.org.my/index.php/development/education/347-tv-retards-your-childs-development, accessed July 17, 2016.
  Also See:
  "More arguments against TV" Consumers Association of Penang. https://consumer.org.my/more-arguments-against-tv/, March 18, 2022.

227

- Sharon Jayson. “Generation Y’s goal? Wealth and fame” *USA Today*. http://usatoday30.usatoday.com/news/nation/2007-01-09-gen-y-cover_x.htm, Posted January 9,2007, Updated January 10, 2007.

228

- Pengcheng Wang; Jia Nie; Xingchao Wang; Yuhui Wang; Fengqing Zhao; Xiaochun Xie; Li Lei; Mingkun Ouyang,. “How are smartphones associated with adolescent materialism?” Journal of Health Psychology Vol. 25 (September 19, 2018). https://journals.sagepub.com/doi/10.1177/1359105318801069.

229

- Tim Kasser. *The High Price of Materialism*. Bradford Books. August 29, 2003.
- Christine Carter. “Three Risks of Too Much Screen Time for Teens” *Greater Good.* Greater Good Science Center (GGSC), University of California, Berkeley. https://greatergood.berkeley.edu/article/item/three_risks_of_too_much_screen_time_for_teens, November 27, 2018.

230

- “TV retards your child’s development” *Consumers Association of Penang.* http://www.consumer.org.my/index.php/development/education/347-tv-retards-your-childs-development, accessed July 17, 2016.
  Also See:
  “More arguments against TV” Consumers Association of Penang. https://consumer.org.my/more-arguments-against-tv/, March 18, 2022.
- Emma Henderson. “Watching lots of TV ‘makes you stupid’” *The Independent.* http://www.independent.co.uk/news/science/watching-lots-of-tv-makes-you-stupid-says-american-universities-a6759026.html, December 3, 2015.
- Herbert E. Krugman and Eugene L. Hartley. “Passive Learning From Television” *The Public Opinion Quarterly,* Vol. 34, No. 2. (Oxford University Press, 1970) pp. 184-190.
- Herbert E. Krugman. “Brain Wave Measures of Media Involvement” *How Advertising Works: The Role of Research.* (New York SAGE Publications, 1998) pp. 139–151.

- "Your brain waves change when you watch TV" *I Am Awake.* http://www.iamawake.co/your-brain-waves-change-when-you-watch-tv/, October 11, 2013.
- Kristina Birdsong. "This is Your Child's Brain on TV" *Scientific Learning.* http://www.scilearn.com/blog/how-television-impacts-learning, March 22, 2016.
  http://54.186.226.228/blog/your-childs-brain-tv, March 22, 2016.
- Douglas Fields. "Watching TV Alters Children's Brain Structure and Lowers IQ".
  http://rdouglasfields.com/2015/05/watching-tv-alters-childrens-brain-structure-and-lowers-iq/, May 4, 2015.
- Jamil Zaki. "What, Me Care? Young Are Less Empathetic" *Scientific American*. A recent study finds a decline in empathy among young people in the U.S.
  http://www.scientificamerican.com/article/what-me-care/, December 23, 2010.
- "Children who watch 'excessive' amounts of TV are more likely to have criminal convictions, exhibit aggression and experience negative emotions: study." *New York Daily News*.
  http://www.nydailynews.com/life-style/health/kids-watch-excessive-tv-criminal-convictions-young-adulthood-study-article-1.1267868, February 19, 2013.
  "With every hour in front of the television, kids were more likely to show aggressive behavior or receive a criminal conviction by early adulthood, according to a study published in 'Pediatrics.' The issue isn't necessarily the content of the programming, but the social isolation that comes from so many hours in front of the tube."
- David L. Hill. "Why to Avoid TV Before Age 2" (early brain development). *American Academy of Pediatrics*.
  http://www.healthychildren.org/English/family-life/Media/Pages/Why-to-Avoid-TV-Before-Age-2.aspx, May 11, 2013.
  Original Source: Dad to Dad: Parenting Like a Pro (Copyright © American Academy of Pediatrics 2012).
- Alice Park. "Baby Einsteins: Not So Smart After All" *Time*.
  http://content.time.com/time/health/article/0,8599,1650352,00.html August 06, 2007.

- "Television vs. Reading" *Parent Soup®*, a Trademark of iVillage℠ Inc. Copyright 1996. http://webshare.northseattle.edu/fam180/topics/tv/tvvsread.htm, accessed May 29, 2014.
- Jim Trelease. *The Read-Aloud Handbook*, Penguin Books, 1995.
- Marie Winn. "Television and the Brain" and "Brain Changes" *Plug-In Drug.* (New York: Viking Penguin, 1977, Revised and Updated Edition, 2002) pp. 67–69:
  "'Research conducted during the next two decades removed any doubt about the impact of early brain stimulation on a child's later cognitive development. ... they were able to demonstrate that environmental factors can alter neuron pathways during early childhood and long after ... among the most important of the environmental factors ... are the language and eye contact an infant is exposed to [and] ... the number of words an infant hears each day is the single most important predictor of later intelligence, school success and social competence. But there's one catch. The words have to come from an attentive, engaged human being ... radio and television do not work."

231

- Herbert E. Krugman and Eugene L. Hartley. "Passive Learning From Television" *The Public Opinion Quarterly,* Vol. 34, No. 2. (Oxford University Press, 1970) pp. 184-190.
- Herbert E. Krugman. "Brain Wave Measures of Media Involvement" *How Advertising Works: The Role of Research.* (New York SAGE Publications, 1998) pp. 139–151.
- "Your brain waves change when you watch TV" *I Am Awake.* http://www.iamawake.co/your-brain-waves-change-when-you-watch-tv/, October 11, 2013.

232

- Richard Freed. "The Tech Industry's War on Kids: How psychology is being used as a weapon against children" *Blog, Child and adolescent psychologist* Richard Freed. https://medium.com/@richardnfreed/the-tech-industrys-psychological-war-on-kids-c452870464ce, March 12, 2018.

233

- Richard Freed. “The Tech Industry’s War on Kids: How psychology is being used as a weapon against children” *Blog, Child and adolescent psychologist* Richard Freed. https://medium.com/@richardnfreed/the-tech-industrys-psychological-war-on-kids-c452870464ce, March 12, 2018.

234

- Nellie Bowles. “The Digital Gap Between Rich and Poor Kids Is Not What We Expected” *The New York Times.* https://www.nytimes.com/2018/10/26/style/digital-divide-screens-schools.html, October. 26, 2018.

235

- Herbert E. Krugman and Eugene L. Hartley. “Passive Learning From Television” *The Public Opinion Quarterly,* Vol. 34, No. 2. (Oxford University Press, 1970) pp. 184-190.
- Herbert E. Krugman. “Brain Wave Measures of Media Involvement” *How Advertising Works: The Role of Research.* (New York SAGE Publications, 1998) pp. 139–151.
- “Your brain waves change when you watch TV” *I Am Awake.* http://www.iamawake.co/your-brain-waves-change-when-you-watch-tv/, October 11, 2013.
- Allan Stromfeldt Christensen. “Lemminged: to be herded off the peak oil cliff by filmmakers” *TransitionVoice.com.* http://transitionvoice.com/2014/11/lemminged-to-be-herded-off-the-peak-oil-cliff-by-filmmakers/, November 11, 2014.
- “Television: Opiate of the Masses” *FamilyResource.com.* http://www.familyresource.com/lifestyles/mental-environment/television-opiate-of-the-masses, accessed May 29, 2014.
- Douglas Fields. “Watching TV Alters Children’s Brain Structure and Lowers IQ”. http://rdouglasfields.com/2015/05/watching-tv-alters-childrens-brain-structure-and-lowers-iq/, May 4, 2015.
- Kristina Birdsong. “This is Your Child’s Brain on TV” *Scientific Learning.* http://www.scilearn.com/blog/how-television-impacts-learning, March 22, 2016. http://54.186.226.228/blog/your-childs-brain-tv, March 22, 2016.

236

- Tom Nichols. *The Death of Expertise.* p. 143. New York: Oxford University Press, 2017.

237

- Robert Kubey and Mihaly Csikszentmihalyi. "Television Addiction is no mere metaphor" *Scientific American.* http://www.academia.edu/5065840/Television_Addiction_is_no_mere_metaphor, accessed January 23, 2016.

238

- Christopher Ingraham. "Today's men are not nearly as strong as their dads were, researchers say" *Washington Post.* https://www.washingtonpost.com/news/wonk/wp/2016/08/15/todays-men-are-nowhere-near-as-strong-as-their-dads-were-researchers-say/, August 15, 2016.

239

- Melissa Dahl. "How Neuroscientists Explain the Mind-Clearing Magic of Running" Science of Us, *Huffington Post.* https://www.huffingtonpost.com/science-of-us/how-neuroscientists-expla_b_9787466.html, December 6, 2017.

240

- University of Tsukuba. "Active body, active mind: The secret to a younger brain may lie in exercising your body" *ScienceDaily.* https://www.sciencedaily.com/releases/2015/10/151023084456.htm, October 23, 2015.

241

- Travis Bradberry and Jean Greaves. *Emotional Intelligence 2.0.* p. 132. San Diego, California: TalentSmart, 2009.

242

- "Want a younger brain? Stay in school — and take the stairs" *Science Daily.* https://www.sciencedaily.com/releases/2016/03/160309125520.htm March 9, 2016.

243

- Celestine Chua. "Top 10 Reasons You Should Stop Watching TV" *Personal Excellence: Be Your Best Self, Live Your Best Life.* http://personalexcellence.co/blog/top-10-reasons-you-should-stop-watching-tv/, May 2, 2010.

244

- "Children who watch 'excessive' amounts of TV are more likely to have criminal convictions, exhibit aggression and experience negative emotions: study." *New York Daily News.* http://www.nydailynews.com/life-style/health/kids-watch-excessive-tv-criminal-convictions-young-adulthood-study-article-1.1267868, February 19, 2013.
  "With every hour in front of the television, kids were more likely to show aggressive behavior or receive a criminal conviction by early adulthood, according to a study published in 'Pediatrics.' The issue isn't necessarily the content of the programming, but the social isolation that comes from so many hours in front of the tube."

245

- Travis Bradberry. "13 Things Mentally Strong People Won't Do". https://www.linkedin.com/pulse/13-things-mentally-strong-people-wont-do-dr-travis-bradberry/, September 11, 2017.

246

- Jean M. Twenge, W. Keith Campbell. "Associations between screen time and lower psychological well-being among children and adolescents: Evidence from a population-based study" *Preventive Medicine Reports,* Volume 12, Pages 271-283. Elsevier B.V. https://doi.org/10.1016/j.pmedr.2018.10.003, October 18, 2018.

247

- Travis Bradberry and Jean Greaves. *Emotional Intelligence 2.0.* p. 196. San Diego, California: TalentSmart, 2009.

248

- Travis Bradberry and Jean Greaves. *Emotional Intelligence 2.0.* p. 237. San Diego, California: TalentSmart, 2009.

249

- Victoria L. Dunckley, MD. "Screentime and Arrested Social Development" *Psychology Today.* https://www.psychologytoday.com/us/blog/mental-wealth/201606/screentime-and-arrested-social-development, June 30, 2016.

250

- Victoria L. Dunckley, MD. "Screentime and Arrested Social Development" *Psychology Today.* https://www.psychologytoday.com/us/blog/mental-wealth/201606/screentime-and-arrested-social-development, June 30, 2016.

251

- Travis Bradberry and Jean Greaves. *Emotional Intelligence 2.0.* pp. 52, 53. San Diego, California: TalentSmart, 2009.

252

- Travis Bradberry. "Eight Habits of Considerate People". https://www.linkedin.com/pulse/eight-habits-considerate-people-dr-travis-bradberry, November 8, 2017.

253

- Travis Bradberry and Jean Greaves. *Emotional Intelligence 2.0.* pp. 244, 245. San Diego, California: TalentSmart, 2009.

254

- Travis Bradberry. "13 Things Mentally Strong People Won't Do". https://www.linkedin.com/pulse/13-things-mentally-strong-people-wont-do-dr-travis-bradberry/, September 11, 2017.

255

- Travis Bradberry and Jean Greaves. *Emotional Intelligence 2.0.* p. 245. San Diego, California: TalentSmart, 2009.

256

- Travis Bradberry and Jean Greaves. *Emotional Intelligence 2.0.* p. 75. San Diego, California: TalentSmart, 2009.

257

- Travis Bradberry. "These are the habits that mentally strong people rely on" *World Economic Forum.* https://www.weforum.org/agenda/2016/10/habits-to-help-you-develop-mental-strength, October 26, 2016.

258

- Travis Bradberry. "9 Habits of Profoundly Influential People". https://www.linkedin.com/pulse/critical-habits-profoundly-influential-people-dr-travis-bradberry/, July 20, 2015.

259

- Travis Bradberry. "12 Habits of Genuine People". https://www.linkedin.com/pulse/importance-being-genuine-dr-travis-bradberry/, November 15, 2015.

260

- Leslie Becker-Phelps. "Don't Just React: Choose Your Response" *Psychology Today.* https://www.psychologytoday.com/blog/making-change/201307/dont-just-react-choose-your-response, July 23, 2013.

261

- Leslie Becker-Phelps. "Don't Just React: Choose Your Response" *Psychology Today.* https://www.psychologytoday.com/blog/making-change/201307/dont-just-react-choose-your-response, July 23, 2013.

262

- "Children who watch 'excessive' amounts of TV are more likely to have criminal convictions, exhibit aggression and experience negative emotions: study" *New York Daily News.* http://www.nydailynews.com/life-style/health/kids-watch-excessive-tv-criminal-convictions-young-adulthood-study-article-1.1267868, February 19, 2013.

- Douglas Fields. “Watching TV Alters Children’s Brain Structure and Lowers IQ”. http://rdouglasfields.com/2015/05/watching-tv-alters-childrens-brain-structure-and-lowers-iq/, May 4, 2015.

263

- Jamil Zaki. “What, Me Care? Young Are Less Empathetic” *Scientific American*. A recent study finds a decline in empathy among young people in the U.S. http://www.scientificamerican.com/article/what-me-care/, December 23, 2010.
- “Children who watch ‘excessive’ amounts of TV are more likely to have criminal convictions, exhibit aggression and experience negative emotions: study.” *New York Daily News*. http://www.nydailynews.com/life-style/health/kids-watch-excessive-tv-criminal-convictions-young-adulthood-study-article-1.1267868, February 19, 2013.
  “With every hour in front of the television, kids were more likely to show aggressive behavior or receive a criminal conviction by early adulthood, according to a study published in ‘Pediatrics.’ The issue isn’t necessarily the content of the programming, but the social isolation that comes from so many hours in front of the tube.”
- “TV retards your child’s development” *Consumers Association of Penang*. http://www.consumer.org.my/index.php/development/education/347-tv-retards-your-childs-development, accessed July 17, 2016.
  Also See:
  “More arguments against TV” Consumers Association of Penang. https://consumer.org.my/more-arguments-against-tv/, March 18, 2022.

264

- Bevan Hurley. “Family claim 10-year-old obsessed with gadgets fatally shot his mother for refusing to buy him a VR headset” *The Independent*. https://www.independent.co.uk/news/world/americas/crime/quiana-mann-son-murder-virtual-reality-b2253013.html, December 29, 2022.

265

- Nathan Place. "Son accused of murdering parents after pretending to work at SpaceX" *The Independent*. https://www.independent.co.uk/news/world/americas/crime/son-murder-parents-spacex-wisconsin-b1988292.html, January 7, 2022.

266

- "ISU study finds TV viewing, video game play contribute to kids' attention problems" *Iowa State University*. http://www.news.iastate.edu/news/2010/jul/TVVGattention, July 4, 2010.

267

- Eileen Kennedy-Moore. "Do Boys Need Rough and Tumble Play?" *Psychology Today*. https://www.psychologytoday.com/blog/growing-friendships/201506/do-boys-need-rough-and-tumble-play, June 30, 2015.
- Jaak Panksepp. "Can PLAY diminish ADHD and facilitate the construction of the social brain?" *Journal of the Canadian Academy of Child and Adolescent Psychiatry* (16, 57-66), 2007.

268

- "Children who watch 'excessive' amounts of TV are more likely to have criminal convictions, exhibit aggression and experience negative emotions: study." *New York Daily News*. http://www.nydailynews.com/life-style/health/kids-watch-excessive-tv-criminal-convictions-young-adulthood-study-article-1.1267868, February 19, 2013.
  "With every hour in front of the television, kids were more likely to show aggressive behavior or receive a criminal conviction by early adulthood, according to a study published in 'Pediatrics.' The issue isn't necessarily the content of the programming, but the social isolation that comes from so many hours in front of the tube."

269

- Sergio M. Pellis and Vivien C. Pellis. "Rough-and-Tumble Play: Training and Using the Social Brain" *The Oxford Handbook of the Development of Play* (December 2010).
  Online Reference:
  http://www.oxfordhandbooks.com/view/10.1093/oxfordhb/9780195393002.001.0001/oxfordhb-9780195393002-e-019, September 2012.
- Eileen Kennedy-Moore. "Do Boys Need Rough and Tumble Play?" *Psychology Today*.
  https://www.psychologytoday.com/blog/growing-friendships/201506/do-boys-need-rough-and-tumble-play, June 30, 2015.

270

- "Masculine and Feminine Roles in Relationships (Sociology Essay)" UKEssays.
  http://www.ukessays.com/essays/sociology/masculine-and-feminine-roles-in-relationships-sociology-essay.php, November 2013, updates May 2017, November 2018. "Since 2010 ... it is evident that in the modern family set ups a large number of men are taking up the feminine roles."
- Hugh Wilson. "Are men becoming more feminine?" *MSN.com*.
  http://him.uk.msn.com/in-the-know/are-men-becoming-more-feminine-women-male-female-gender-gap, July 19, 2013.
- Anderson-Niles, Amanda. "Gender Switch: When Did Men Become So Feminine?" Urban Belle Magazine.
  https://urbanbellemag.com/2010/10/gender-switch-when-did-men-become-so-feminine/, October 18, 2010.
- Christopher Ingraham. "Today's men are not nearly as strong as their dads were, researchers say" *Washington Post*.
  https://www.washingtonpost.com/news/wonk/wp/2016/08/15/todays-men-are-nowhere-near-as-strong-as-their-dads-were-researchers-say/, August 15, 2016.

271

- Eileen Kennedy-Moore. "Do Boys Need Rough and Tumble Play?" *Psychology Today*.
  https://www.psychologytoday.com/blog/growing-friendships/201506/do-boys-need-rough-and-tumble-play, June 30, 2015.

- Peter K. Smith. "Chapter 6: Physical Activity Play: Exercise Play and Rough-and-Tumble" *Children and Play* (April 2009).
  Online Reference:
  https://onlinelibrary.wiley.com/doi/10.1002/9781444311006.ch6, April 17, 2009.

272

- Stuart Brown. "Play, Spirit, and Character" *Interview for* On Being with Krista Tippett*, a Peabody Award-winning public radio show and podcast.* https://onbeing.org/programs/stuart-brown-play-spirit-and-character/, June 19, 2014.

273

- "Masculine and Feminine Roles in Relationships (Sociology Essay)" UKEssays.
  http://www.ukessays.com/essays/sociology/masculine-and-feminine-roles-in-relationships-sociology-essay.php, November 2013, updates May 2017, November 2018. "Since 2010 ... it is evident that in the modern family set ups a large number of men are taking up the feminine roles."
- Hugh Wilson. "Are men becoming more feminine?" *MSN.com*. http://him.uk.msn.com/in-the-know/are-men-becoming-more-feminine-women-male-female-gender-gap, July 19, 2013.
- Anderson-Niles, Amanda. "Gender Switch: When Did Men Become So Feminine?" Urban Belle Magazine.
  https://urbanbellemag.com/2010/10/gender-switch-when-did-men-become-so-feminine/, October 18, 2010.

- Christopher Ingraham. "Today's men are not nearly as strong as their dads were, researchers say" *Washington Post*.
  https://www.washingtonpost.com/news/wonk/wp/2016/08/15/todays-men-are-nowhere-near-as-strong-as-their-dads-were-researchers-say/, August 15, 2016.

274

- "Physical Fighting by Youth" *Child Trends Databank*.
  https://www.childtrends.org/indicators/physical-fighting-by-youth/, Accessed December 23, 2017.

- Anderson-Niles, Amanda. "Gender Switch: When Did Men Become So Feminine?" Urban Belle Magazine. https://urbanbellemag.com/2010/10/gender-switch-when-did-men-become-so-feminine/, October 18, 2010.
- Hugh Wilson. "Are men becoming more feminine?" *MSN.com*. http://him.uk.msn.com/in-the-know/are-men-becoming-more-feminine-women-male-female-gender-gap, July 19, 2013.

275

- "Physical Fighting by Youth" *Child Trends Databank*. https://www.childtrends.org/indicators/physical-fighting-by-youth/, Accessed December 23, 2017.

276

- "Children who watch 'excessive' amounts of TV are more likely to have criminal convictions, exhibit aggression and experience negative emotions: study." *New York Daily News*. http://www.nydailynews.com/life-style/health/kids-watch-excessive-tv-criminal-convictions-young-adulthood-study-article-1.1267868, February 19, 2013.
  "With every hour in front of the television, kids were more likely to show aggressive behavior or receive a criminal conviction by early adulthood, according to a study published in 'Pediatrics.' The issue isn't necessarily the content of the programming, but the social isolation that comes from so many hours in front of the tube."

277

- Sheri Madigan, PhD; Dillon Browne, PhD; Nicole Racine, PhD. "Association Between Screen Time and Children's Performance on a Developmental Screening Test" *American Medical Association: JAMA Pediatr.* 2019;173(3):244-250. doi:10.1001/jamapediatrics.2018.5056. https://jamanetwork.com/journals/jamapediatrics/fullarticle/2722666, January 28, 2019.
- Chloe Reichel. "The health effects of screen time on children: A research roundup" *Journalist's Resource*. https://journalistsresource.org/studies/society/public-health/screen-time-children-health-research/, May 14, 2019.

278

- Matt Richtel. “A Silicon Valley School That Doesn’t Compute” *The New York Times.* https://www.nytimes.com/2011/10/23/technology/at-waldorf-school-in-silicon-valley-technology-can-wait.html, October 22, 2011.

279

- Owenz, Meghan. “The Rich Get Smart, The Poor Get Technology: The New Digital Divide in School Choice”*Screenfreeparenting.com*. https://www.screenfreeparenting.com/rich-get-smart-poor-get-technology-new-digital-divide-school-choice/, November 21, 2017.

280

- Michael Park; Erin Leahey; Russell J. Funk. “Papers and patents are becoming less disruptive over time” *Nature* 613, 138–144 (2023). https://doi.org/10.1038/s41586-022-05543-x (https://www.nature.com/articles/s41586-022-05543-x), January 4, 2023.

281

- Michael Park; Erin Leahey; Russell J. Funk. “Papers and patents are becoming less disruptive over time” *Nature* 613, 138–144 (2023). https://doi.org/10.1038/s41586-022-05543-x (https://www.nature.com/articles/s41586-022-05543-x), January 4, 2023.

282

- Michael Park; Erin Leahey; Russell J. Funk. “Papers and patents are becoming less disruptive over time” *Nature* 613, 138–144 (2023). https://doi.org/10.1038/s41586-022-05543-x (https://www.nature.com/articles/s41586-022-05543-x), January 4, 2023.

283

- Jane M. Healy. *Endangered Minds: Why Children Don’t Think And What We Can Do About It*. Simon & Schuster. October 15, 1999.
  Jane M. Healy. “Endangered Minds” Creating the Future: Perspectives on Educational Change. Ed. Dee Dickinson. Johns Hopkins University. 2012.

284

- Tom Nichols. *The Death of Expertise.* pp. 119–122, 138. New York: Oxford University Press, 2017.
- Matthew Fisher et al. "Searching for Explanations: How the Internet Inflates Estimates of Internal Knowledge" *Journal of Experimental Psychology* 144(3). pp. 674–687, June 2015.
- Tom Jacobs. "Searching the Internet Creates an Illusion of Knowledge" *Pacific Standard online.* https://psmag.com/environment/searching-internet-creates-the-illusion-of-knowledge-, April 1, 2015.
- Richard Arum. "College Graduates: Satisfied, but Adrift" in *The State of the American Mind*. p. 73, Mark Bauerlein and Adam Bellow, eds. West Conshohocken, PA: Templeton, 2015.

285

- Thomas Friedman. *Thank You for Being Late.* p. 378. New York: Farrar, Straus and Giroux (Picador), 2017.

286

- Thomas Friedman. *Thank You for Being Late.* p. 378. New York: Farrar, Straus and Giroux (Picador), 2017.

287

- Maryanne Wolf. "Skim reading is the new normal. The effect on society is profound" *The Guardian*. https://www.theguardian.com/commentisfree/2018/aug/25/skim-reading-new-normal-maryanne-wolf, August 25, 2018.

288

- "ISU study finds TV viewing, video game play contribute to kids' attention problems" *Iowa State University*. http://www.news.iastate.edu/news/2010/jul/TVVGattention, July 4, 2010.

289

- Martin Hickman. "Watching TV 'makes toddlers less intelligent'" *The Independent.* http://www.independent.co.uk/news/education/education-news/watching-tv-makes-toddlers-less-intelligent-1960856.html, May 2, 2010.

290

- David Hinckley. "Average American watches 5 hours of TV per day, report shows" *New York Daily News*. https://www.nydailynews.com/life-style/average-american-watches-5-hours-tv-day-article-1.1711954, March 5, 2014.
- Nielsen's data showing average weekly usage for ascending age groups. Ages:
    - 2–11: 24 hours, 16 minutes. (round down to 24 hours, 15 minutes, or 24.25 x 10 years counting ages 2 and 11 = 12,610)
    - 12–17: 20 hours, 41 minutes. (round down to 20 hours, 30 minutes, or 20.5 hours x 6 years counting ages 12 and 17 = 6,396)
    - 18–24: 22 hours, 27 minutes. (round down to 22 hours, 20 minutes, or 20.33 hours x 6 years counting age 18 to end of age 23 = 6,343)
    - By age 24, the average American has watched 25,349 hours of TV (does not include PC/Internet video time). Here is the breakdown continuing older age groups, showing people watch more and more as they get older. Ages:
        - 25–34: 27 hours, 36 minutes.
        - 35–49: 33 hours, 40 minutes.
        - 50–64: 43 hours, 56 minutes.
        - 65–plus: 50 hours, 34 minutes.
- Victor C. Strasburger, MD, FAAP, and Marjorie J. Hogan, MD, FAAP. "Children, Adolescents, and the Media" *American Academy of Pediatrics*. http://pediatrics.aappublications.org/content/132/5/958.full, May 11, 2013.

291

- "TV retards your child's development" *Consumers Association of Penang*. http://www.consumer.org.my/index.php/development/education/347-tv-retards-your-childs-development, accessed July 17, 2016.
  Also See:
  "More arguments against TV" Consumers Association of Penang. https://consumer.org.my/more-arguments-against-tv/, March 18, 2022.

- Emma Henderson. “Watching lots of TV ‘makes you stupid’” *The Independent.* http://www.independent.co.uk/news/science/watching-lots-of-tv-makes-you-stupid-says-american-universities-a6759026.html, December 3, 2015.
- Herbert E. Krugman and Eugene L. Hartley. “Passive Learning From Television” *The Public Opinion Quarterly,* Vol. 34, No. 2. (Oxford University Press, 1970) pp. 184-190.
- Herbert E. Krugman. “Brain Wave Measures of Media Involvement” *How Advertising Works: The Role of Research.* (New York SAGE Publications, 1998) pp. 139–151.
- “Your brain waves change when you watch TV” *I Am Awake.* http://www.iamawake.co/your-brain-waves-change-when-you-watch-tv/, October 11, 2013.
- Kristina Birdsong. “This is Your Child’s Brain on TV” *Scientific Learning.* http://www.scilearn.com/blog/how-television-impacts-learning, March 22, 2016.
  http://54.186.226.228/blog/your-childs-brain-tv, March 22, 2016.
- Douglas Fields. “Watching TV Alters Children’s Brain Structure and Lowers IQ”. http://rdouglasfields.com/2015/05/watching-tv-alters-childrens-brain-structure-and-lowers-iq/, May 4, 2015.
- Jamil Zaki. “What, Me Care? Young Are Less Empathetic” *Scientific American*. A recent study finds a decline in empathy among young people in the U.S. http://www.scientificamerican.com/article/what-me-care/, December 23, 2010.
- “Children who watch ‘excessive’ amounts of TV are more likely to have criminal convictions, exhibit aggression and experience negative emotions: study.” *New York Daily News*. http://www.nydailynews.com/life-style/health/kids-watch-excessive-tv-criminal-convictions-young-adulthood-study-article-1.1267868, February 19, 2013.
  “With every hour in front of the television, kids were more likely to show aggressive behavior or receive a criminal conviction by early adulthood, according to a study published in ‘Pediatrics.’ The issue isn’t necessarily the content of the programming, but the social isolation that comes from so many hours in front of the tube.”

- David L. Hill. "Why to Avoid TV Before Age 2" (early brain development). *American Academy of Pediatrics*. http://www.healthychildren.org/English/family-life/Media/Pages/Why-to-Avoid-TV-Before-Age-2.aspx, May 11, 2013.
Original Source: Dad to Dad: Parenting Like a Pro (Copyright © American Academy of Pediatrics 2012).
- Alice Park. "Baby Einsteins: Not So Smart After All" *Time*. http://content.time.com/time/health/article/0,8599,1650352,00.html August 06, 2007.
- "Television vs. Reading" *Parent Soup*®, a Trademark of iVillage℠ Inc. Copyright 1996. http://webshare.northseattle.edu/fam180/topics/tv/tvvsread.htm, accessed May 29, 2014.
- Jim Trelease. *The Read-Aloud Handbook*, Penguin Books, 1995.
- Edward L. Schor. *Caring for your school-age child : ages 5 to 12.* American Academy of Pediatrics. New York : Bantam Books, 2004.
"What Children are NOT Doing When Watching TV" American Academy of Pediatrics. http://www.healthychildren.org/English/family-life/Media/Pages/What-Children-are-NOT-Doing-When-Watching-TV.aspx, May 11, 2013.
Also See: *Caring for Our Children*. A Joint Collaborative Project of the American Academy of Pediatrics and the American Public Health Association. https://nrckids.org/files/CFOC4%20pdf-%20FINAL.pdf, 2019.
- Tina D. Hoang, MSPH; Jared Reis, PhD; Na Zhu, MD, MPH; David R. Jacobs Jr, PhD; Lenore J. Launer, PhD; Rachel A. Whitmer, PhD; Stephen Sidney, MD; Kristine Yaffe, MD.
"Effect of Early Adult Patterns of Physical Activity and Television Viewing on Midlife Cognitive Function" *JAMA Psychiatry* (The Journal of the American Medical Association). http://archpsyc.jamanetwork.com/article.aspx?articleid=2471270, January 2016, Vol 73, No. 1.
- Marie Winn. "Television and the Brain," "Brain Changes," "Losing the Thread," "The Basic Building Blocks," "A Preference for Watching," "Free Time and Resourcefulness" *Plug-In Drug.* (New York: Viking Penguin, 1977, Revised and Updated Edition, 2002) pp. 67–69, 95–99, 131.

292

- Ana Tintocalis. "San Francisco Middle Schools No Longer Teaching 'Algebra 1'" *The California Report*, KQED News. https://www.kqed.org/news/10610214/san-francisco-middle-schools-no-longer-teaching-algebra-1, July 22, 2015.

293

- Kaelan Deese. "Oregon governor signs bill ending reading and math proficiency requirements for graduation" *YahooNews*. https://www.yahoo.com/now/oregon-governor-signs-bill-ending-154100667.html, August 10, 2021.

294

- Sarah Randazzo. "Schools Ditch Homework, Deadlines" *The Wall Street Journal*, Vol. CCLXXXI No. 97. Thursday April 27, 2023.

295

- "Top 10 Best Clark County Public Schools (2023)" *Nevada Public School Review*. https://www.publicschoolreview.com/nevada/clark-county, accessed April 28, 2023.

296

- Jay Mathews. "Some kids need harder lessons than schools are willing to give them: New data reveals stubborn preference for below grade-level instruction" *The Washington Post*. https://www.washingtonpost.com/education/2022/08/21/grade-level-reading-difficulty-challenge/, August 21, 2022.
- TNTP. "Unlocking Acceleration: How Below Grade-Level Work is Holding Students Back in Literacy" *The New Teacher Project* (tntp.org). https://tntp.org/assets/documents/Unlocking_Acceleration_8.16.22.pdf, August 2022.

297

- TNTP. "Unlocking Acceleration: How Below Grade-Level Work is Holding Students Back in Literacy" *The New Teacher Project* (tntp.org). https://tntp.org/assets/documents/Unlocking_Acceleration_8.16.22.pdf, August 2022.
- Jay Mathews. "Some kids need harder lessons than schools are willing to give them: New data reveals stubborn preference for below grade-level instruction" *The Washington Post*. https://www.washingtonpost.com/education/2022/08/21/grade-level-reading-difficulty-challenge/, August 21, 2022.

298

- TNTP. "Accelerate, Don't Remediate: New Evidence from Elementary Math Classrooms" *The New Teacher Project* (tntp.org). https://tntp.org/publications/view/teacher-training-and-classroom-practice/accelerate-dont-remediate, https://tntp.org/assets/documents/TNTP_Accelerate_Dont_Remediate_FINAL.pdf, May 23, 2021.
- TNTP. "Unlocking Acceleration: How Below Grade-Level Work is Holding Students Back in Literacy" *The New Teacher Project* (tntp.org). https://tntp.org/assets/documents/Unlocking_Acceleration_8.16.22.pdf, August 2022.
- Jay Mathews. "Some kids need harder lessons than schools are willing to give them: New data reveals stubborn preference for below grade-level instruction" *The Washington Post*. https://www.washingtonpost.com/education/2022/08/21/grade-level-reading-difficulty-challenge/, August 21, 2022.

299

- Douglas Belkin, Ben Chapma, and Ben Kesling. "'How Do I Do That?' The New Hires of 2023 Are Unprepared for Work" *The Wall Street Journal*. https://www.wsj.com/articles/lost-learning-remote-pandemic-workplace-skills-new-employees-51351b33.

300

- Douglas Belkin, Ben Chapma, and Ben Kesling. "'How Do I Do That?' The New Hires of 2023 Are Unprepared for Work" *The Wall Street Journal.* https://www.wsj.com/articles/lost-learning-remote-pandemic-workplace-skills-new-employees-51351b33.

301

- Douglas Belkin, Ben Chapma, and Ben Kesling. "'How Do I Do That?' The New Hires of 2023 Are Unprepared for Work" *The Wall Street Journal.* https://www.wsj.com/articles/lost-learning-remote-pandemic-workplace-skills-new-employees-51351b33.

302

- Douglas Belkin, Ben Chapma, and Ben Kesling. "'How Do I Do That?' The New Hires of 2023 Are Unprepared for Work" *The Wall Street Journal.* https://www.wsj.com/articles/lost-learning-remote-pandemic-workplace-skills-new-employees-51351b33.

303

- Douglas Belkin, Ben Chapma, and Ben Kesling. "'How Do I Do That?' The New Hires of 2023 Are Unprepared for Work" *The Wall Street Journal.* https://www.wsj.com/articles/lost-learning-remote-pandemic-workplace-skills-new-employees-51351b33.

304

- Douglas Belkin, Ben Chapma, and Ben Kesling. "'How Do I Do That?' The New Hires of 2023 Are Unprepared for Work" *The Wall Street Journal.* https://www.wsj.com/articles/lost-learning-remote-pandemic-workplace-skills-new-employees-51351b33.

305

- Catherine Rampell. "The rise of the 'gentleman's A' and the GPA arms race" *The Washington Post.* https://www.washingtonpost.com/opinions/the-rise-of-the-gentlemans-a-and-the-gpa-arms-race/2016/03/28/05c9e966-f522-11e5-9804-537defcc3cf6_story.html, March 28, 2016.

306

- Emma Henderson. “Watching lots of TV ‘makes you stupid’” *The Independent.* http://www.independent.co.uk/news/science/watching-lots-of-tv-makes-you-stupid-says-american-universities-a6759026.html, December 3, 2015.
- Tina D. Hoang, MSPH; Jared Reis, PhD; Na Zhu, MD, MPH; David R. Jacobs Jr, PhD; Lenore J. Launer, PhD; Rachel A. Whitmer, PhD; Stephen Sidney, MD; Kristine Yaffe, MD.
  “Effect of Early Adult Patterns of Physical Activity and Television Viewing on Midlife Cognitive Function” *JAMA Psychiatry* (The Journal of the American Medical Association).
  http://archpsyc.jamanetwork.com/article.aspx?articleid=2471270, January 2016, Vol 73, No. 1.

307

- Catherine Rampell. “The rise of the ‘gentleman’s A’ and the GPA arms race” *The Washington Post.* https://www.washingtonpost.com/opinions/the-rise-of-the-gentlemans-a-and-the-gpa-arms-race/2016/03/28/05c9e966-f522-11e5-9804-537defcc3cf6_story.html, March 28, 2016.

308

- Ray Williams. “Anti-Intellectualism and the ‘Dumbing Down’ of America” Indiana University Center for Civic Literacy. https://blogs.iu.edu/civicliteracy/2015/05/22/anti-intellectualism-and-the-dumbing-down-of-america/, July 7, 2014, Online May 22, 2015.

309

- Educational Testing Service (ETS). “America’s Skills Challenge: Millennials and the Future” *The ETS Center for Research on Human Capital and Education.* Princeton, NJ: January 2015.
  https://www.ets.org/s/research/30079/asc-millennials-and-the-future.pdf.
  Note: The Educational Testing Service (ETS) is the academic organization that administers the SAT tests to high school students bound for college.

310

- Kristina Birdsong. “This is Your Child’s Brain on TV” *Scientific Learning.* http://www.scilearn.com/blog/how-television-impacts-learning, March 22, 2016.
  http://54.186.226.228/blog/your-childs-brain-tv, March 22, 2016.

311

- Jean M. Twenge. "Have Smartphones Destroyed a Generation?" *The Atlantic*. https://www.theatlantic.com/magazine/archive/2017/09/has-the-smartphone-destroyed-a-generation/534198/, September 2017.

312

- Kevin Carey. "Americans Think We Have the World's Best Colleges. We Don't." *The New York Times.* https://www.nytimes.com/2014/06/29/upshot/americans-think-we-have-the-worlds-best-colleges-we-dont.html, June 28, 2014.

313

- Tom Nichols. *The Death of Expertise.* p. 93. New York: Oxford University Press, 2017.

314

- Tom Nichols. *The Death of Expertise.* p. 93. New York: Oxford University Press, 2017.

315

- Catherine Rampell. "The rise of the 'gentleman's A' and the GPA arms race" *The Washington Post.* https://www.washingtonpost.com/opinions/the-rise-of-the-gentlemans-a-and-the-gpa-arms-race/2016/03/28/05c9e966-f522-11e5-9804-537defcc3cf6_story.html, March 28, 2016.

316

- Tom Nichols. *The Death of Expertise.* pp. 74–75. New York: Oxford University Press, 2017.

317

- Jeremy Warner. "Harsh truths about the decline of Britain" *The Telegraph.* http://www.telegraph.co.uk/finance/economics/10417838/Harsh-truths-about-the-decline-of-Britain.html, October 31, 2013.

318

- Alice Cuddy. "The IQ of Europeans is dropping due to technology, say researchers" *Euronews.* https://www.euronews.com/2017/12/29/the-iq-of-europeans-is-dropping-due-to-technology-say-researchers, December 29, 2017.
- James Flynn, Michael Shayer. "IQ decline and Piaget: Does the rot start at the top?" *Intelligence,* Volume 66, Pages 112-121. Elsevier B.V. https://doi.org/10.1016/j.intell.2017.11.010, December 8, 2017.

319

- Tom Nichols. *The Death of Expertise.* p. 99. New York: Oxford University Press, 2017.

320

- Greg Toppo. "Americans trail adults in other countries in math, literacy, problem-solving" *USA Today.*
  "New international study finds U.S. workers lag in math, reading, problem-solving"
  "Adult literacy scores in 12 countries higher than USA, only five score lower"
  http://www.usatoday.com/story/news/nation/2013/10/08/literacy-international-workers-education-math-americans/2935909/, October 8, 2013

- Eric Berger. "By 2010 most science Ph.D.s will go to foreign-born students" *SciGuy.* http://blog.chron.com/sciguy/2007/11/by-2010-most-science-ph-d-s-will-go-to-foreign-born-students/, November 21, 2007.

321

- Elizabeth Redden. "Foreign Student Dependence" *Inside Higher Education.* https://www.insidehighered.com/news/2013/07/12/new-report-shows-dependence-us-graduate-programs-foreign-students, July 12, 2013.

322

- Lisa Rapaport. "Screen time linked to ADHD symptoms in teens" *Reuters.* https://www.reuters.com/article/us-health-adhd-digital-media/screen-time-linked-to-adhd-symptoms-in-teens-idUSKBN1K72L8, July 17, 2018.

323

- Jeff Haden. *The Motivation Myth.* p. 16. New York: Penguin, 2018.

324

- Jeff Haden. *The Motivation Myth.* p. 15. New York: Penguin, 2018.

325

- "Home-Schooling: Outstanding results on national tests" *The Washington Times.* http://www.washingtontimes.com/news/2009/aug/30/home-schooling-outstanding-results-national-tests/, August 30, 2009.

326

- Tom Nichols. *The Death of Expertise.* pp. 72–76. New York: Oxford University Press, 2017.

327

- Tom Nichols. *The Death of Expertise.* p. 94. New York: Oxford University Press, 2017.

328

- Tom Nichols. *The Death of Expertise.* p. 71. New York: Oxford University Press, 2017.

329

- Tom Nichols. *The Death of Expertise.* p. 72. New York: Oxford University Press, 2017.

330

- Scott Jaschik. "Let the Right Ones In" *Slate.com.* http://www.slate.com/articles/life/inside_higher_ed/2014/10/college_admissions_rose_hulman_institute_of_technology_uses_locus_of_control.html, October 30, 2014.

331

- Travis Bradberry. "9 Surprising Things Ultra Productive People Do Every Day" https://www.linkedin.com/pulse/surprising-things-ultra-productive-people-do-every-day-bradberry/, November 7, 2016.

332

- Cynthia Kubu, PhD, and Andre Machado, MD. "The Science Is Clear: Why Multitasking Doesn't Work" *Cleveland Clinic Health Essentials*. https://health.clevelandclinic.org/science-clear-multitasking-doesnt-work/, June 1, 2017, updated March 10, 2021.

333

- Travis Bradberry and Jean Greaves. *Emotional Intelligence 2.0.* p. 147. San Diego, California: TalentSmart, 2009.

334

- Matt Richtel. "The Myth of Multitasking" *Attached to Technology and Paying a Price*, p. 2. *NYTimes.com*.
  Updated Title: "Your Brain on Computers: Attached to Technology and Paying a Price"
  https://www.nytimes.com/2010/06/07/technology/07brain.html, June 6, 2010.
- Johanna Rothman. "Why Multitasking Doesn't Work" *Pragmatic Manager*. http://www.jrothman.com/2011/01/why-multitasking-doesnt-work/, January 1, 2011.
- Porter Anderson. "Study: Multitasking is counterproductive" *CNN*. https://www.cnn.com/2001/CAREER/trends/08/05/multitasking.study/index.html, December 6, 2001.

335

- Joseph Stromberg. "Why you should take notes by hand — not on a laptop". *Vox: Science & Health*. https://www.vox.com/2014/6/4/5776804/note-taking-by-hand-versus-laptop, March 31, 2015.

336

- Anne Sliper Midling, Norwegian University of Science and Technology. "Why writing by hand makes kids smarter: Writing by hand creates much more activity in the sensorimotor parts of the brain, researchers found." *ScienceDaily*. https://www.sciencedaily.com/releases/2020/10/201001113540.htm, October 1, 2020.

- Eva Ose Askvik, PhD; F. R. (Ruud) van der Weel, PhD; Audrey L. H. van der Meer, PhD. “The Importance of Cursive Handwriting Over Typewriting for Learning in the Classroom: A High-Density EEG Study of 12-Year-Old Children and Young Adults.” *Frontiers in Psychology*. https://doi.org/10.3389/fpsyg.2020.01810 https://www.frontiersin.org/articles/10.3389/fpsyg.2020.01810/full, July 28, 2020.

337

- Steinar Brandslet. “Students learn better writing by hand: We learn much better if we take notes the old-fashioned way, by hand, and now researchers know more about why this is.” *Futurity*. Original Study DOI: 10.3389/fpsyg.2023.1219945. https://www.futurity.org/writing-by-hand-learning-3253352-2/, October 15, 2024.

- F. R. (Ruud) van der Weel, PhD; Audrey L. H. van der Meer, PhD. “Handwriting but not typewriting leads to widespread brain connectivity: a high-density EEG study with implications for the classroom.” *Frontiers in Psychology, Sec. Educational Psychology*, Volume 14 - 2023. https://doi.org/10.3389/fpsyg.2023.1219945 https://www.frontiersin.org/journals/psychology/articles/10.3389/fpsyg.2023.1219945/full, January 25, 2024.

338

- Perri Klass, M.D. “Why Handwriting Is Still Essential in the Keyboard Age” *The New York Times*. https://well.blogs.nytimes.com/2016/06/20/why-handwriting-is-still-essential-in-the-keyboard-age, June 20, 2016.

339

- Perri Klass, M.D. “Why Handwriting Is Still Essential in the Keyboard Age” *The New York Times*. https://well.blogs.nytimes.com/2016/06/20/why-handwriting-is-still-essential-in-the-keyboard-age, June 20, 2016.

340

- Maryanne Wolf. “Skim reading is the new normal. The effect on society is profound” *The Guardian*. https://www.theguardian.com/commentisfree/2018/aug/25/skim-reading-new-normal-maryanne-wolf, August 25, 2018.

341

- Patricia Alexander and Lauren Singer Trakhman. “The enduring power of print for learning in a digital world” *The Conversation*. https://theconversation.com/the-enduring-power-of-print-for-learning-in-a-digital-world-84352, October 3, 2017.
- Patricia Alexander and Lauren Singer Trakhman. “Reading Across Mediums: Effects of Reading Digital and Print Texts on Comprehension and Calibration” *The Journal of Experimental Education*, Pages 155-172. Published online: March 9, 2016. https://doi.org/10.1080/00220973.2016.1143794

342

- Naomi S. Baron. “Why do we remember more by reading in print vs. on a screen?” *Neuropsych*. https://bigthink.com/neuropsych/reading-memory/, May 9, 2021.

- Naomi S. Baron. *How We Read Now: Strategic Choices for Print, Screen, and Audio*. New York: Oxford University Press. March 22, 2021.

- Anne Mangen; Bente R. Walgermo; Kolbjørn Brønnick. “Reading linear texts on paper versus computer screen: Effects on reading comprehension” *International Journal of Educational Research*, Volume 58, 2013, Pages 61–68. https://doi.org/10.1016/j.ijer.2012.12.002, January 5, 2013.

343

- Ceridwen Dovey. “Can Reading Make You Happier?” *The New Yorker*. http://www.newyorker.com/culture/cultural-comment/can-reading-make-you-happier, June 9, 2015.

344

- Thu-Huong Ha. "New research links reading books with longer life" *Quartz*. http://qz.com/754109/new-research-links-reading-books-with-longer-life/, August 10, 2016.

345

- Danny Heitman. "I'm Revisiting the Books of My Youth" *The Wall Street Journal*, Vol. CCLXXXI No. 97. Thursday April 27, 2023.

346

- Amanda Gardner. "TV watching raises risk of health problems, dying young" *CNN*. http://www.cnn.com/2011/HEALTH/06/14/tv.watching.unhealthy/, June 14, 2011.
    - "For every two hours Americans spend watching TV each day, there are 176 new cases of diabetes, 38 additional deaths from heart disease, and 104 additional deaths due to any cause per 100,000 people per year"—That's 2 hours/day per 100,000 people. That means given the US population of 324,000,000, and the fact that average US TV viewing is 4 hours per day—TV viewing causes a total of 1,140,480 new cases of diabetes, 246,240 additional deaths from heart disease, and 673,920 additional deaths, every year.

347

- EJ Fox and Mike Spies. "Who Was America's Most Well-Spoken President?" *Vocativ*. http://www.vocativ.com/interactive/usa/us-politics/presidential-readability/, October 10, 2014.
  and
  Fairchild, Quentin B. "Presidential speechmaking in an age of ignorance" News-Press. https://www.news-press.com/story/opinion/contributors/2014/11/05/pr

348

- Stephen Knott, ed. "Life Before the Presidency" *American President: George Washington (1732–1799), Essays on George Washington and His Administration*. Miller Center, University of Virginia. http://millercenter.org/president/washington/essays/biography/2, accessed November 19, 2014.

- Lindsay M. Chervinsky. “George Washington: Life in Brief” Miller Center, University of Virginia. https://millercenter.org/president/washington, Accessed March 18, 2022 https://millercenter.org/president/washington/life-in-brief, Accessed March 18, 2022.

349

- EJ Fox and Mike Spies. “Who Was America’s Most Well-Spoken President?” *Vocativ*. http://www.vocativ.com/interactive/usa/us-politics/presidential-readability/, October 10, 2014.

350

- EJ Fox and Mike Spies. “Who Was America’s Most Well-Spoken President?” *Vocativ*. http://www.vocativ.com/interactive/usa/us-politics/presidential-readability/, October 10, 2014.
  and
  Fairchild, Quentin B. “Presidential speechmaking in an age of ignorance” News-Press. https://www.news-press.com/story/opinion/contributors/2014/11/05/pr

351

- Caroline McCarthy. “Hulu: We’re evil, and proud of it” *CNet*. https://www.cnet.com/news/hulu-were-evil-and-proud-of-it/, February 2, 2009.
  The example video: https://youtu.be/Ek8T4F1ZHG8.

352

- Victoria L. Dunckley, MD. “Screentime and Arrested Social Development” *Psychology Today*. https://www.psychologytoday.com/us/blog/mental-wealth/201606/screentime-and-arrested-social-development, June 30, 2016.

353

- Rosalina Richards, Rob McGee, Sheila M Williams, David Welch, and Robert J Hancox. "Adolescent Screen Time and Attachment to Parents and Peers" *Archives of Pediatrics & Adolescent Medicine.* 164, no. 3 (March 2010): 258–62. doi:10.1001/archpediatrics.2009.280. https://jamanetwork.com/journals/jamapediatrics/fullarticle/382905, March 1, 2010.

354

- Jean M. Twenge. "Have Smartphones Destroyed a Generation?" *The Atlantic.* https://www.theatlantic.com/magazine/archive/2017/09/has-the-smartphone-destroyed-a-generation/534198/, September 2017.

355

- Amy Joyce. "How helicopter parents are ruining college students" *The Washington Post.* http://www.washingtonpost.com/news/parenting/wp/2014/09/02/how-helicopter-parents-are-ruining-college-students/, September 2, 2014.

356

- Marie Winn. "Waiting on Children" *Plug-In Drug.* (New York: Viking Penguin, 1977, Revised and Updated Edition, 2002) p. 146.

357

- Sharon Jayson. "Generation Y's goal? Wealth and fame" *USA Today.* http://usatoday30.usatoday.com/news/nation/2007-01-09-gen-y-cover_x.htm, Posted January 9,2007, Updated January 10, 2007.

358

- Amy Joyce. "How helicopter parents are ruining college students" The Washington Post. http://www.washingtonpost.com/news/parenting/wp/2014/09/02/how-helicopter-parents-are-ruining-college-students/, September 2, 2014.

359

- Jean M. Twenge. "Have Smartphones Destroyed a Generation?" *The Atlantic.* https://www.theatlantic.com/magazine/archive/2017/09/has-the-smartphone-destroyed-a-generation/534198/, September 2017.

360

- Dan Kopf. "The share of American young adults living with their parents is the highest in 75 years" *Quartz.* https://qz.com/1248081/the-share-of-americans-age-25-29-living-with-parents-is-the-highest-in-75-years/, April 10, 2018.

361

- "Adulting" *Dictionary.com.* https://www.dictionary.com/e/slang/adulting/, accessed December 8, 2022.

362

- "Word of the Year 2016: Shortlist" *Oxford Languages, Oxford University Press.* https://languages.oup.com/word-of-the-year/2016-shortlist/, accessed December 8, 2022.

363

- "Adulting" *Dictionary.com.* https://www.dictionary.com/e/slang/adulting/, accessed December 8, 2022.

364

- "Adulting" *Urban Dictionary.* https://www.urbandictionary.com/define.php?term=Adulting, accessed December 8, 2022.

365

- Katy Steinmetz. "This Is What 'Adulting' Means" *Time Magazine.* https://time.com/4361866/adulting-definition-meaning/, June 8, 2016.

366

- "Oxford Word of the Year 2022" *Oxford Languages, Oxford University Press.* https://languages.oup.com/word-of-the-year/2022/, accessed December 8, 2022.

367

- Juliana Kim. "How 'goblin mode' became Oxford's word of the year" *NPR.* https://www.npr.org/2022/12/05/1140696560/oxford-word-2022-goblin-mode, December 5, 2022.

368

- Allan Stromfeldt Christensen. "Lemminged: to be herded off the peak oil cliff by filmmakers" *TransitionVoice.com.* http://transitionvoice.com/2014/11/lemminged-to-be-herded-off-the-peak-oil-cliff-by-filmmakers/, November 11, 2014.

369

- Bob Sornson. "Who's Looking Out for the Children?" *Early Learning Foundation.* http://earlylearningfoundation.com/whos-looking-out-for-the-children/, accessed March 18, 2017.

370

- "Television Addiction" *All About Life Challenges*. http://www.allaboutlifechallenges.org/television-addiction.htm, accessed September 21, 2014.
- "Television: Opiate of the Masses" *FamilyResource.com*. http://www.familyresource.com/lifestyles/mental-environment/television-opiate-of-the-masses, accessed May 29, 2014.

371

- Celestine Chua. "Top 10 Reasons You Should Stop Watching TV" *Personal Excellence: Be Your Best Self, Live Your Best Life.* http://personalexcellence.co/blog/top-10-reasons-you-should-stop-watching-tv/, May 2, 2010.

372

- Jeff Haden. *The Motivation Myth.* p. 168. New York: Penguin, 2018.

373

- Travis Bradberry and Jean Greaves. *Emotional Intelligence 2.0.* p. 69. San Diego, California: TalentSmart, 2009.

374

- Jeff Haden. *The Motivation Myth.* p. 17. New York: Penguin, 2018.

375

- Travis Bradberry and Jean Greaves. *Emotional Intelligence 2.0.* pp. 119. San Diego, California: TalentSmart, 2009.

376

- Travis Bradberry. "13 Things Mentally Strong People Won't Do". https://www.linkedin.com/pulse/13-things-mentally-strong-people-wont-do-dr-travis-bradberry/, September 11, 2017.

377

- Jim Taylor. "Parenting: The Sad Misuse of Self-esteem" *Psychology Today.* https://www.psychologytoday.com/blog/the-power-prime/201002/parenting-the-sad-misuse-self-esteem, February 22, 2010.

378

- Travis Bradberry. "These are the habits that mentally strong people rely on" *World Economic Forum.* https://www.weforum.org/agenda/2016/10/habits-to-help-you-develop-mental-strength, October 26, 2016.

379

- Travis Bradberry. "10 Things Mentally Strong People Won't Do". https://www.forbes.com/sites/travisbradberry/2016/09/20/10-things-mentally-strong-people-wont-do/, September 20, 2016.

380

- Jeff Haden. *The Motivation Myth.* pp. 28–29. New York: Penguin, 2018.

381

- Peter M. Gollwitzer et al., "When Intentions Go Public: Does Social Reality Widen the Intention-Behavior Gap?" *Psychological Science* 20, no. 5 (May 1, 2009): 612. http://journals.sagepub.com/doi/abs/10.1111/j.1467-9280.2009.02336.x, May 1, 2009.
- Jeff Haden. *The Motivation Myth.* p. 28. New York: Penguin, 2018.

382

- Peter M. Gollwitzer et al., "When Intentions Go Public: Does Social Reality Widen the Intention-Behavior Gap?" *Psychological Science* 20, no. 5 (May 1, 2009): 612. http://journals.sagepub.com/doi/abs/10.1111/j.1467-9280.2009.02336.x, May 1, 2009.

383

- Travis Bradberry and Jean Greaves. *Emotional Intelligence 2.0.* p. 68. San Diego, California: TalentSmart, 2009.

384

- Jim Taylor. "Parenting: The Sad Misuse of Self-esteem" *Psychology Today.* https://www.psychologytoday.com/blog/the-power-prime/201002/parenting-the-sad-misuse-self-esteem, February 22, 2010.

385

- Travis Bradberry. "13 Things Mentally Strong People Won't Do". https://www.linkedin.com/pulse/13-things-mentally-strong-people-wont-do-dr-travis-bradberry/, September 11, 2017.

386

- Travis Bradberry. "12 Habits of Genuine People". https://www.linkedin.com/pulse/importance-being-genuine-dr-travis-bradberry/, November 15, 2015.

387

- Travis Bradberry. "13 Things Mentally Strong People Won't Do". https://www.linkedin.com/pulse/13-things-mentally-strong-people-wont-do-dr-travis-bradberry/, September 11, 2017.

388

- William Frierson. "Dream vs. Reality: What Happens After Graduation" *College Recruiter*. https://www.collegerecruiter.com/blog/2015/07/14/dream-vs-reality-what-happens-after-graduation/, July 14, 2015.

389

- Carrie Courogen. "9 Harsh Realities About Graduating College That I Wish Someone Had Warned Me About" *Bustle*. https://www.bustle.com/articles/80461-9-harsh-realities-about-graduating-college-that-i-wish-someone-had-warned-me-about, May 4, 2015.

390

- Travis Bradberry. "13 Things Mentally Strong People Won't Do". https://www.linkedin.com/pulse/13-things-mentally-strong-people-wont-do-dr-travis-bradberry/, September 11, 2017.

391

- Jeff Haden. *The Motivation Myth.* p. 236. New York: Penguin, 2018.

392

- Thomas Friedman. *Thank You for Being Late.* p. 219. New York: Farrar, Straus and Giroux (Picador), 2017.

393

- Thomas Friedman. *Thank You for Being Late.* p. 95. New York: Farrar, Straus and Giroux (Picador), 2017.

394

- Thomas Friedman. *Thank You for Being Late.* p. 226. New York: Farrar, Straus and Giroux (Picador), 2017.

395

- Thomas Friedman. *Thank You for Being Late.* p. 232. New York: Farrar, Straus and Giroux (Picador), 2017.

396

- Thomas Friedman. *Thank You for Being Late.* p. 219. New York: Farrar, Straus and Giroux (Picador), 2017.

397

- Jeff Haden. *The Motivation Myth.* p. 180. New York: Penguin, 2018.

398

- Jeff Haden. *The Motivation Myth.* p. 9. New York: Penguin, 2018.

399

- Thomas Friedman. *Thank You for Being Late.* p. 244. New York: Farrar, Straus and Giroux (Picador), 2017.

400

- Thomas Friedman. *Thank You for Being Late.* p. 245. New York: Farrar, Straus and Giroux (Picador), 2017.

401

- Travis Bradberry. "These are the habits that mentally strong people rely on" *World Economic Forum.* https://www.weforum.org/agenda/2016/10/habits-to-help-you-develop-mental-strength, October 26, 2016.

402

- Travis Bradberry. "11 Things Ultra-Productive People Do Differently" *Forbes.* https://www.forbes.com/sites/travisbradberry/2015/05/13/11-things-ultra-productive-people-do-differently/, May 13, 2015.

403

- Jeff Haden. *The Motivation Myth.* p. 132–135, 140. New York: Penguin, 2018.

404

- Travis Bradberry. "These are the habits that mentally strong people rely on" *World Economic Forum.* https://www.weforum.org/agenda/2016/10/habits-to-help-you-develop-mental-strength, October 26, 2016.

405

- Travis Bradberry. "These are the habits that mentally strong people rely on" *World Economic Forum.* https://www.weforum.org/agenda/2016/10/habits-to-help-you-develop-mental-strength, October 26, 2016.

406

- Travis Bradberry. "These are the habits that mentally strong people rely on" *World Economic Forum.* https://www.weforum.org/agenda/2016/10/habits-to-help-you-develop-mental-strength, October 26, 2016.

407

- Travis Bradberry. "These are the habits that mentally strong people rely on" *World Economic Forum.* https://www.weforum.org/agenda/2016/10/habits-to-help-you-develop-mental-strength, October 26, 2016.

408

- Travis Bradberry. "13 Things Mentally Strong People Won't Do". https://www.linkedin.com/pulse/13-things-mentally-strong-people-wont-do-dr-travis-bradberry/, September 11, 2017.

409

- Jeff Haden. *The Motivation Myth.* p. 32. New York: Penguin, 2018.

410

- Travis Bradberry. "These are the habits that mentally strong people rely on" *World Economic Forum.* https://www.weforum.org/agenda/2016/10/habits-to-help-you-develop-mental-strength, October 26, 2016.
- W. Mischel and E. B. Ebbesen (1970). "Attention in delay of gratification" *Journal of Personality and Social Psychology*, 16(2), 329–337. https://doi.org/10.1037/h0029815.

411

- Jeff Haden. *The Motivation Myth.* p. 37. New York: Penguin, 2018.

412

- Jeff Haden. *The Motivation Myth.* p. 136. New York: Penguin, 2018.

413

- Travis Bradberry. "13 Things Mentally Strong People Won't Do". https://www.linkedin.com/pulse/13-things-mentally-strong-people-wont-do-dr-travis-bradberry/, September 11, 2017.

414

- Dick Van Dyke. *My Lucky Life*. (New York: Random House, 2011) p. 184.

415

- "Fred Astaire Dancer (1899–1987)" *The Biography.com website*. https://www.biography.com/people/fred-astaire-9190991, Last Updated, April 27, 2017.

416

- Travis Bradberry. "13 Things Mentally Strong People Won't Do". https://www.linkedin.com/pulse/13-things-mentally-strong-people-wont-do-dr-travis-bradberry/, September 11, 2017.

417

- Travis Bradberry. "12 Habits of Genuine People". https://www.linkedin.com/pulse/importance-being-genuine-dr-travis-bradberry/, November 15, 2015.

418

- Travis Bradberry. "Eight Habits of Considerate People". https://www.linkedin.com/pulse/eight-habits-considerate-people-dr-travis-bradberry, November 8, 2017.

419

- *Vincent van Gogh Gallery*. http://www.vggallery.com/, accessed July, 30 2015.
- *Heilbrunn Timeline of Art History*. The Metropolitan Museum of Art. http://www.metmuseum.org/toah/hd/gogh/hd_gogh.htm, accessed July, 30 2015.

420

- Henry James. *The Tragic Muse.* p. 582. London: Rupert Hart-Davis, 1948.

421

- Travis Bradberry and Jean Greaves. *Emotional Intelligence 2.0.* p. 188. San Diego, California: TalentSmart, 2009.

422

- Ryan J. Dwyer, Kostadin Kushlev, Elizabeth W. Dunn. "Smartphone use undermines enjoyment of face-to-face social interactions" *Journal of Experimental Social Psychology.* Elsevier B.V. https://www.sciencedirect.com/science/article/pii/S0022103117301737, September 2018.

423

- Adrian F. Ward, Kristen Duke, Ayelet Gneezy, and Maarten W. Bos. "Brain Drain: The Mere Presence of One's Own Smartphone Reduces Available Cognitive Capacity". *Journal of the Association for Consumer Research* 2, no. 2 (April 2017): 140-154. https://doi.org/10.1086/691462, April 3, 2017.

424

- Travis Bradberry. "Eight Habits of Considerate People". https://www.linkedin.com/pulse/eight-habits-considerate-people-dr-travis-bradberry, November 8, 2017.

425

- Dan Michel, *Remorse of Conscience*, or *Ayenbite of inwyt*, ed. Richard Morris (London: N. Trubner & Co., 1895) pp. 74-75. Morris' 1895 republication is from Michel's 1340 translation from French to Kentish. Michel's 1340 translation is of the 13th century French *Somme le Roi*. My quote is from a translation of the 1340 Kentish into modern English, translation by Judith G. Humphries, in "The Personification of Death in Middle English Literature" (Denton, Texas: 1970, North Texas State University) p. 10.

426

- Travis Bradberry. “Eight Habits of Considerate People”. https://www.linkedin.com/pulse/eight-habits-considerate-people-dr-travis-bradberry, November 8, 2017.

427

- Travis Bradberry. “13 Habits of Exceptionally Likeable People”. https://www.linkedin.com/pulse/13-habits-exceptionally-likeable-people-dr-travis-bradberry, January 27, 2015.

428

- Travis Bradberry. “13 Habits of Exceptionally Likeable People”. https://www.linkedin.com/pulse/13-habits-exceptionally-likeable-people-dr-travis-bradberry, January 27, 2015.

429

- Travis Bradberry and Jean Greaves. *Emotional Intelligence 2.0.* pp. 160, 161. San Diego, California: TalentSmart, 2009.

430

- Richard Alan Krieger. *Civilization’s Quotations: Life’s Ideal.* (New York: Algora Publishing, 2002) p. 122.

431

- Benjamin Disraeli, Earl of Beaconsfield, quoted in *Puck*, Volume XV, No. 370 (p. 95). April 9, 1884. Google digitized version: https://books.google.com/books?id=0_kiAQAAMAAJ&pg=PA95.

432

- Aristotle. *Goodreads*. http://www.goodreads.com/quotes/31240-happiness-is-a-state-of-activity, July 24, 2015.

433

- Bruno S. Frey, Christine Benesch, Alois Stutzer. "Does watching TV make us happy?" *Journal of Economic Psychology*, Volume 28, Issue 3, Pages 283–313, June 2007. Elsevier B.V. (ScienceDirect.com). https://www.bsfrey.ch/wp-content/uploads/2021/08/does-watching-tv-make-us-happy.pdf, February 14, 2007.
  Updated URLs:
  https://doi.org/10.1016/j.joep.2007.02.001,
  https://www.sciencedirect.com/science/article/abs/pii/S0167487007000104.

434

- Jean Twenge. "Changes in how we're spending our free time might explain the unhappiness epidemic" *PsyPost.org*. https://www.psypost.org/2018/08/changes-in-how-were-spending-our-free-time-might-explain-the-unhappiness-epidemic-51893, August 5, 2018.

435

- Jean Twenge. "Changes in how we're spending our free time might explain the unhappiness epidemic" *PsyPost.org*. https://www.psypost.org/2018/08/changes-in-how-were-spending-our-free-time-might-explain-the-unhappiness-epidemic-51893, August 5, 2018.

436

- Jean Twenge. "Changes in how we're spending our free time might explain the unhappiness epidemic" *PsyPost.org*. https://www.psypost.org/2018/08/changes-in-how-were-spending-our-free-time-might-explain-the-unhappiness-epidemic-51893, August 5, 2018.

437

- Aya Hatano, PhD, Kyoto University; Cansu Ogulmus, PhD, University of Tübingen; Kou Murayama, PhD, University of Tübingen; Hiroaki Shigemasu, PhD, Kochi University of Technology. "Thinking About Thinking: People Underestimate How Enjoyable and Engaging Just Waiting Is" *Journal of Experimental Psychology: General*. https://www.apa.org/news/press/releases/2022/07/thoughts-mind-wander, July 28, 2022.

438

- Travis Bradberry and Jean Greaves. *Emotional Intelligence 2.0.* p. 124. San Diego, California: TalentSmart, 2009.

439

- Shawn Achor. "The Happy Secret to Better Work" *Ted Talk*. February, 2012. https://www.ted.com/talks/shawn_achor_the_happy_secret_to_better_work/transcript?language=en.

440

- Martha Washington. *Letter to Mercy Warren*. 1789. https://en.wikiquote.org/wiki/Martha_Washington.

441

- Leslie Becker-Phelps. "Don't Just React: Choose Your Response" *Psychology Today.* https://www.psychologytoday.com/blog/making-change/201307/dont-just-react-choose-your-response, July 23, 2013.

442

- Franklin Delano Roosevelt. *AZQuotes*. http://www.azquotes.com/quote/250883.

443

- Jeff Haden. *The Motivation Myth.* p. 112. New York: Penguin, 2018.

444

- Travis Bradberry and Jean Greaves. *Emotional Intelligence 2.0.* pp. 51, 52. San Diego, California: TalentSmart, 2009.

445

- Travis Bradberry and Jean Greaves. *Emotional Intelligence 2.0.* pp. 7, 8. San Diego, California: TalentSmart, 2009.
- Shawn Achor. "The Happy Secret to Better Work" *Ted Talk*. February, 2012. https://www.ted.com/talks/shawn_achor_the_happy_secret_to_better_work/transcript?language=en.

446

- Lisa Quast. "Why Grit Is More Important Than IQ When You're Trying To Become Successful" *Forbes.* https://www.forbes.com/sites/lisaquast/2017/03/06/why-grit-is-more-important-than-iq-when-youre-trying-to-become-successful/, March 6, 2017.

447

- Shawn Achor. "The Happy Secret to Better Work" *Ted Talk*. February, 2012. https://www.ted.com/talks/shawn_achor_the_happy_secret_to_better_work/transcript?language=en.

448

- Jon Morrow. "How to Be Smart in a World of Dumb Bloggers" *BoostBlogTraffic*. http://boostblogtraffic.com/smart-blogger/, September 17, 2013. https://smartblogger.com/how-to-start-a-blog/, updated February 28, 2022.
- Jon Morrow. "On Gluttony, Selfishness, and Unleashing the Power Within" *BoostBlogTraffic*. http://boostblogtraffic.com/unleash-your-power/, November 27, 2014.

449

- Jon Morrow. "How to Be Smart in a World of Dumb Bloggers" *BoostBlogTraffic*. http://boostblogtraffic.com/smart-blogger/, September 17, 2013. https://smartblogger.com/how-to-start-a-blog/, updated February 28, 2022.

450

- Jeff Haden. "How the Smartest Minds Use Intuition and Emotional Intelligence to Make Better Decisions, Backed by Science: Experience matters, but research shows high emotional intelligence can also make intuitive decisions even more accurate" *Inc.com*. https://www.inc.com/jeff-haden/how-smartest-minds-use-intuition-emotional-intelligence-to-make-better-decisions-backed-by-science.html, April 8, 2022.

451

- Jeff Haden. "How the Smartest Minds Use Intuition and Emotional Intelligence to Make Better Decisions, Backed by Science: Experience matters, but research shows high emotional intelligence can also make intuitive decisions even more accurate" *Inc.com*. https://www.inc.com/jeff-haden/how-smartest-minds-use-intuition-emotional-intelligence-to-make-better-decisions-backed-by-science.html, April 8, 2022.

452

- Travis Bradberry and Jean Greaves. *Emotional Intelligence 2.0.* p. 61. San Diego, California: TalentSmart, 2009.

453

- Travis Bradberry. "These are the habits that mentally strong people rely on" *World Economic Forum.* https://www.weforum.org/agenda/2016/10/habits-to-help-you-develop-mental-strength, October 26, 2016.

454

- Jeff Haden. "How the Smartest Minds Use Intuition and Emotional Intelligence to Make Better Decisions, Backed by Science: Experience matters, but research shows high emotional intelligence can also make intuitive decisions even more accurate" *Inc.com*. https://www.inc.com/jeff-haden/how-smartest-minds-use-intuition-emotional-intelligence-to-make-better-decisions-backed-by-science.html, April 8, 2022.

455

- Jeff Haden. "How the Smartest Minds Use Intuition and Emotional Intelligence to Make Better Decisions, Backed by Science: Experience matters, but research shows high emotional intelligence can also make intuitive decisions even more accurate" *Inc.com*. https://www.inc.com/jeff-haden/how-smartest-minds-use-intuition-emotional-intelligence-to-make-better-decisions-backed-by-science.html, April 8, 2022.

- Friederike Fabritius; Hans W. Hagemann. *The Leading Brain: Neuroscience Hacks to Work Smarter, Better, Happier*. TarcherPerigee, February 20, 2018.

456

- Dick Van Dyke. *My Lucky Life*. (New York: Random House, 2011) pp. 21–22.

457

- Peter Falk. *Just One More Thing*. (New York: Carroll & Graf, 2007) pp. 165–166.

458

- Blake Bailey. *Philip Roth: The Biography*. (W. W. Norton & Company; 1st Edition [Hardcover], 2021) p. 334.

459

- Blake Bailey. *Philip Roth: The Biography*. (W. W. Norton & Company; 1st Edition [Hardcover], 2021) p. 751.

460

- Marcus Tullius Cicero. *De Oratore,* Book II §36.
  Paraphrase "Historia magistra vitae est" commonly extrapolated from the quote "Historia vero testis temporum, lux veritatis, vita memoriae, magistra vitae, nuntia vetustatis, qua voce alia, nisi oratoris, immortalitati commendatur" The Loeb Classical Library, English translation Cambridge, Massachusetts, Harvard University Press, 1967.
  https://archive.org/stream/cicerodeoratore01ciceuoft/cicerodeoratore01ciceuoft_djvu.txt, 55 BC.

461

- Jeff Haden. *The Motivation Myth.* p. 137. New York: Penguin, 2018.

462

- Belle Beth Cooper. "How to Be a Success at Everything" *Fast Company.* https://www.fastcompany.com/3056613/how-i-became-a-morning-person-read-more-books-and-learned-, February 12, 2016.

463

- Nielsen. "Percentage of Americans who say they watch too much TV: 49 %" *BLS American Time Use Survey*, A.C. Nielsen Co. http://www.statisticbrain.com/television-watching-statistics/, Date Verified: 12.7.2013.
- Robert Kubey and Mihaly Csikszentmihalyi "Television Addiction Is No Mere Metaphor" *Scientific American*. http://www.academia.edu/5065840/Television_Addiction_is_no_mere_metaphor, January 23, 2016.
  Further Information from the authors on television addiction:
  - Television and the Quality of Life: How Viewing Shapes Everyday Experience. Robert Kubey and Mihaly Csikszentmihalyi. Lawrence Erlbaum Associates, 1990.
  - Television Dependence, Diagnosis, and Prevention. Robert W. Kubey in Tuning in to Young Viewers: Social Science Perspectives on Television. Edited by Tannis M. MacBeth. Sage, 1995.
  - "I'm Addicted to Television": The Personality, Imagination, and TV Watching Patterns of Self-Identified TV Addicts. Robert D. McIlwraith in Journal of Broadcasting and Electronic Media, Vol. 42, No. 3, pages 371--386; Summer 1998.
  - The Limited Capacity Model of Mediated Message Processing. Annie Lang in Journal of Communication, Vol. 50, No. 1, pages 46--70; March 2000.
  - Internet Use and Collegiate Academic Performance Decrements. Robert Kubey, Michael J. Lavin and John R. Barrows in Journal of Communication, Vol. 51, No. 2, pages 366--382; June 2001.
- "Television Addiction" *All About Life Challenges*. http://www.allaboutlifechallenges.org/television-addiction.htm, accessed September 21, 2014.

www.ingramcontent.com/pod-product-compliance
Lightning Source LLC
LaVergne TN
LVHW020659110826
845149LV00012B/2055

* 9 7 8 0 9 9 7 3 2 5 0 6 5 *